The Green Revolution

This book reviews the Green Revolution, starting with its inception and development from the 1940s to the 1970s, and leading to what is commonly referred to as a second Green Revolution in the 2000s. Building on the historical assessment, it draws insights for contemporary policy debates and demonstrates important lessons for the here and now.

'Green Revolution' refers to the technical measures employed to increase food (particularly grain) production, based mainly on improved seed varieties for higher yields and pest resistance. For it to be successful the Green Revolution often required land reform, investments in irrigation and fertilizer supply that were not available to women and marginal farmers.

This book analyses three underlying principles that have guided green revolutions: the political environment in which they were set; how they contributed to both the successes and challenges the Green Revolution continues to face; and the systemic institutional barriers for access to these agricultural production advances, with a focus on how gender relations limit the inclusion of women even when they are the principle cultivators and farm managers. The book draws on experiences in Mexico, India and China, examining government policy, the role of the family farm, and key issues around the inclusion of women. In doing so, this book connects the history of the Green Revolution with contemporary policy debates on the developing world, particularly in relation to Africa and Asia, around foreign aid and agricultural research. It also specifically establishes that greater inclusivity for women and other marginalised farming communities will significantly enhance the effectiveness of these programs.

Interlinking themes of development policy, gender, and agricultural research, this book will be of great interest to students and scholars of agricultural development, food security, and sustainable development, as well as policymakers and practitioners working in international aid and agri-food policies.

Patrick Kilby is a Senior Lecturer in the School of Archaeology and Anthropology, Australian National University, Australia.

Earthscan Food and Agriculture

The Financialization of Agri-Food Systems
Contested Transformations
Edited by Hilde Bjørkhaug, André Magnan and Geoffrey Lawrence

Redesigning the Global Seed Commons
Law and Policy for Agrobiodiversity and Food Security
Christine Frison

Organic Food and Farming in China
Top-down and Bottom-up Ecological Initiatives
Steffanie Scott, Zhenzhong Si, Theresa Schumilas and Aijuan Chen

Farming, Food and Nature
A Sustainable Future for Animals, People and the Environment
Edited by Joyce D'Silva and Carol McKenna

Governing Sustainable Seafood
Peter Oosterveer and Simon Bush

Farming Systems and Food Security in Africa
Priorities for Science and Policy Under Global Change
Edited by John Dixon, Dennis P. Garrity, Jean-Marc Boffa, Timothy Olalekan Williams, Tilahun Amede with Christopher Auricht, Rosemary Lott and George Mburathi

Consumers, Meat and Animal Products
Policies, Regulations and Marketing
Terence J. Centner

For further details please visit the series page on the Routledge website:
http://www.routledge.com/books/series/ECEFA/

The Green Revolution
Narratives of Politics, Technology and Gender

Patrick Kilby

First published 2019
by Routledge
2 Park Square, Milton Park, Abingdon, Oxon OX14 4RN

and by Routledge
52 Vanderbilt Avenue, New York, NY 10017

First issued in paperback 2020

Routledge is an imprint of the Taylor & Francis Group, an informa business

© 2019 Patrick Kilby

The right of Patrick Kilby to be identified as author of this work has been asserted by him in accordance with sections 77 and 78 of the Copyright, Designs and Patents Act 1988.

All rights reserved. No part of this book may be reprinted or reproduced or utilised in any form or by any electronic, mechanical, or other means, now known or hereafter invented, including photocopying and recording, or in any information storage or retrieval system, without permission in writing from the publishers.

Trademark notice: Product or corporate names may be trademarks or registered trademarks, and are used only for identification and explanation without intent to infringe.

British Library Cataloguing-in-Publication Data
A catalogue record for this book is available from the British Library

Library of Congress Cataloging-in-Publication Data
Names: Kilby, Patrick, author.
Title: The green revolution : narratives of politics, technology and gender / Patrick Kilby.
Description: New York : Routledge, 2019. |
Series: Earthscan food and agriculture | Includes bibliographical references and index.
Identifiers: LCCN 2018055573 (print) | LCCN 2019003868 (ebook) | ISBN 9780429200823 (eBook) | ISBN 9780367191603 (hardback)
Subjects: LCSH: Green Revolution. | Agricultural innovations.
Classification: LCC S494.5.I5 (ebook) | LCC S494.5.I5 K55 2019 (print) | DDC 338.1/6--dc23
LC record available at https://lccn.loc.gov/2018055573

ISBN 13: 978-0-367-67021-4 (pbk)
ISBN 13: 978-0-367-19160-3 (hbk)

Typeset in Times New Roman
by Taylor & Francis Books

Contents

Acknowledgements		vi
Preface		vii
1	Introduction and background	1
2	The Green Revolution	6
3	Three Green Revolution case studies	24
4	Countervailing forces: Structural adjustment and the twenty-first century Green Revolution	42
5	The Green Revolution and absent women	56
6	Conclusion	72
Index		74

Acknowledgements

This book has enabled me to draw together three life-long passions: agriculture, history and gender, or more to the point why women have been systematically excluded in a whole range of public policy areas in development, in this case rural development and agriculture. For this I am indebted to the Australian-American Fulbright Commission for having faith in me and providing me with a Fellowship to undertake research in the US. I am also indebted to Professor Vara Prasad and Dr Jan Middendorf of the Sustainable Intensification Innovation Lab in the College of Agriculture at Kansas State University, for providing me opportunity and support to work with them on improving the gender impact and reach of their agriculture research in developing countries under the USAID supported Feed the Future Program. Finally, special thanks go to my wife, friend, and colleague Dr Joyce Wu who is a lifelong inspiration to me and provided valuable insights and input in the preparation of this book

Preface

This book reviews what was known as the Green Revolution in agricultural research from the 1940s to the 1970s, in the context of Cold War politics, using case studies from Mexico, India, and China. This is related by way of background of rapid agricultural change in the late nineteenth century in Europe, the US, Australia, and elsewhere, where the role of the state in relation to smallholder and peasant farmers was important for success. This is then put in the context of what is referred to as a second Green Revolution in the 2000s, which focusses mainly on Africa. A recurring theme is the forgotten women farmers in all of these processes and that these technical advances were and still remain gender blind. This is despite a significant proportion of farmers or farm managers being women who are left out of these research and related processes. Finally, there are two overarching themes central to this narrative: the role of the state in supporting these programs, and how neo-liberal-based ideology can hinder the state; and secondly, the almost religious zeal in the belief that technological advances can solve problems like global hunger and also win not only the Cold War but more recent global rivalries such as with China in Africa.

1 Introduction and background

Introduction

The term Green Revolution has been used to refer to the advanced plant breeding of the 1950s and 1960s, which led to large increases in food grain production, mainly in wheat, rice, and maize in Asia and parts of Latin America. Of course, the story is more complex and is made up of an amalgam of political expediency, technical innovation, and social marginalisation, of which poor women make up the largest group. In this book I will explore these political and social processes over the past 150 years and show that, while technical advances in food production are important for socio-economic development, they can also result in further marginalisation and contribute to growing inequality.

'Green Revolution' is now almost an aphorism to cover the full suite of measures to increase food (particularly grain) production. These include additional measures and complementary policies such as land reform, investments in irrigation and fertilizer supply, and integrated rural development programs. The people who led the Green Revolution in the 1940s and 1950s include the Rockefeller brothers through their Foundation, and US plant breeders, most notably Nobel Peace Prize winner Norman Borlaug. Later the World Bank, and government donors from the US and the West joined in and began funding the work. China at the same time was carrying out its own 'green revolution', elements of which it shared with other developing countries.

In the 2000s, what is referred to as a second Green Revolution is underway to revive the work of the first Green Revolution but with a widened scope to include a broader range of crops, and into geographic areas where the first Green Revolution was less successful such as Africa. I will argue that this second Green Revolution of the twenty-first century has the same structural issues that bedevilled the first Green Revolution. That is, while increasing overall food supply, there has been an increase in equality and social marginalisation in the agricultural sector, particularly that of women farmers, who historically have been absent from any policy or research considerations, and as I will argue in this book, still continue to be overlooked.

The emphasis on the technical advances in crop breeding in the decades immediately following the Second World War masks the structural and policy enablers necessary for the Green Revolution to succeed, and this is the reason for differential impacts within and between countries. The key challenge continues to be how to adapt Green Revolution technologies to local production systems under quite different national and socio-political contexts. Of course, this is not new as Clawson and Hoy (1979) made this point 40 years ago:

> rather than endeavoring to change the peasant to be compatible with the Green Revolution, we can attempt to modify the Green Revolution products to be compatible with the peasant's world. This would automatically reduce risk to the peasant and thereby increase his adoption rate and possibly his level of living.
>
> (p.384)

Developing technologies to the broadest range of farmers (including women peasant farmers) and their farming systems, rather than focusing on male farmers on larger farms, which has largely been the case up until now, is key. The family farm still remains the centre of agricultural production systems, but this has changed and adapted over the last 50 years. The main change is that now, more than ever, the principle cultivator is often the woman, while the man migrates elsewhere for work in order to diversify family incomes.

Women farmers, however, have been continually overlooked, as new technology tends to be developed for larger and more capitalised farms. Corporate farming is often seen as the solution, but it can also be a false economy and lead to social dislocation. This book argues that overlooking women farmers highlights some of the structural and political issues that limit the potential of further agriculture production increases. These issues are important to understand if the Green Revolution of the twenty-first century is to be successful.

This book will review the Green Revolution in its many iterations, focusing particularly on a series of technical and policy innovations that occurred over the past 75 years or more. How they have affected broader socio-economic development outcomes for both men and women farmers, and why they may have been included or excluded will be analysed.

I will start with a brief review of the US led post-war Green Revolution from the 1940s to the 1970s, but also will look at events in the rural sector of Europe, US, Australia and elsewhere, in the latter half of the nineteenth century to provide some context, as they were the antecedents of the Green Revolution and together have also been referred to as a 'green revolution' (Zanden 1991; Harwood 2012). There was also a parallel process that occurred in the Peoples Republic of China (PRC) after the Second World War and, similarly, other activities such as the World Bank investments in large dams for irrigation in the 1950s cannot be ignored. These were important complements to the Green Revolution and were part of the modernisation philosophy of the time, which the Green Revolution embodied. The more recent 'second' Green Revolution, mainly in Africa, and associated programs such as the Millennium Villages Program, and China's agricultural programs there will be looked at in the light of historical antecedents and missed opportunities, particularly by not including women farmers in research policy and practice.

The focus of the book will be on analysing the underlying principles that guided these interventions, the political environment in which they were set, and how they contributed to both the successes and challenges the Green Revolution continues to face. This book will draw together the systemic institutional barriers for access to these agricultural production advances, with a particular focus on women, and how gender relations limit their access even when they are the principle cultivator, responsible for some if not all crops grown on peasant farms.

Background

The post-war Green Revolution is generally seen as having its origins in the Rockefeller brothers', Nelson and John D.'s interest in agricultural development from the late 1920s when the Rockefeller Foundation introduced agricultural research into its Natural Sciences Division (Ekbladh 2002). In the 1940s, following Rockefeller Foundation nascent plant breeding projects in China to produce high yielding varieties (HYVs) in the late 1930s, they introduced plant breeding projects into Mexico in the early 1940s (Perkins 1990). These projects followed a visit by US Vice President Henry Wallace (himself a plant breeder) who drove to Mexico to attend the inauguration of the new President Camacho, and also to rebuild US links with Mexico to protect the US from having either a communist government or a military dictatorship on its border. Wallace sent Nelson Rockefeller to Mexico in 1941 to look for ways to increase food production as a way to stabilise the Mexican government against the risk of a communist insurgency. The other incentive was that it also helped Rockefeller protect his business interests in the country after the nationalisation of his company Standard Oil (Hill 2017; Patel 2013; Perkins 1990).

Nelson Rockefeller then sent a team of plant breeders including Norman Borlaug and George Harrar to Mexico in 1943 to develop high yielding varieties of wheat initially, and later maize and rice. This was arguably the first 'shot' of the post-war Green Revolution that transformed the world and led to an end to famine and chronic food shortages, providing the foundation for regular food surpluses in most Green Revolution countries. What was absent from these 'successes' was an equitable distribution of benefits across rural communities, with a strong bias to the larger more capitalised farmers, and a 'betting on the strong' approach by recipient governments. In many cases there was a decline in nutritional standards as monoculture became the norm in order to achieve secure food supplies at the national level (Negin et al. 2009; Pinstrup-Andersen and Hazell 1985). There was also little attempt to address the complex social relations in communities including recognising women as farmers in their own right or making any attempt to meet their needs. Before looking at these issues in detail it is worth stepping back in time to look at the processes in Europe and the US and Australia of the late nineteenth century, and the central role the small-scale peasant farmer played in that earlier 'Green Revolution's success, and how it provides lessons for more recent Green Revolution research.

References

Clawson, David L., and Don R. Hoy. 1979. "Nealtican, Mexico: A peasant community that rejected the 'Green Revolution'." *American Journal of Economics and Sociology* 38(4): 371–387.

Ekbladh, David. 2002. "'Mr. TVA': Grass-roots development, David Lilienthal, and the rise and fall of the Tennessee Valley Authority as a symbol for US overseas development, 1933–1973." *Diplomatic History* 26(3): 335–374.

Harwood, Jonathan. 2012. *Europe's Green Revolution and Others Since: The Rise and Fall of Peasant-Friendly Plant Breeding*. Abingdon: Routledge.

Hill, Brent. 2017. "The Green Revolution and the Rockefellers." *Isocraccy*, Feb. 28.

Negin, Joel, Roseline Remans, Susan Karuti, and Jessica C. Fanzo. 2009. "Integrating a broader notion of food security and gender empowerment into the African Green Revolution." *Food Security* 1(3): 351–360.

Patel, Raj. 2013. "The long green revolution." *The Journal of Peasant Studies* 40(1): 1–63.
Perkins, John H. 1990. "The Rockefeller Foundation and the green revolution, 1941–1956." *Agriculture and Human Values* 7(3–4): 6–18.
Pinstrup-Andersen, Per, and Peter B.R. Hazell. 1985. "The impact of the Green Revolution and prospects for the future." *Food Reviews International* 1(1): 1–25.
Zanden, J.L. van. 1991. "The first green revolution: The growth of production and productivity in European agriculture, 1870–1914." *The Economic History Review* 44(2): 215–239.

2 The Green Revolution

Introduction

This chapter tracks the history of the Green Revolution and its antecedents in Europe, the US and the West more generally in the late nineteenth century and the first half of the twentieth. It focusses on the role of state support to peasant agriculture and the role of the family farm in those early successes. This is contrasted with the post-war Green Revolution which had a strong technical focus led by the Rockefeller brothers and their Foundation, as part of modernisation principles of the time which included large dams and massive irrigation schemes to feed and power the world that were emblematic of a series of Cold War rivalries. The Green Revolution while painted as the epitome of Western technologic superiority, was paralleled by a very similar set of processes occurring in China at the same time. The important difference though was the focus of China first on peasant agriculture, then on large-scale collective farms that failed, and back to supporting peasant agriculture, all with strong government support. The Green Revolution supported by the West had a focus on larger peasant farmers, a 'betting on the strong', that was at the expense of small-scale peasant and in particular women farmers who have been overlooked in all iterations of the Green Revolution.

The Green Revolution of the late nineteenth early twentieth centuries

While the Green Revolution has been characterised as unique to a certain period and series of events, it was in fact part of a continuum of technological change in agriculture that took place after the industrial revolution of the eighteenth century if not before (Harwood 2012; Zanden 1991). It was in response to social and economic changes and the consequent demands for increased food availability by an increasingly urban population. Zanden (1991) argues that the first Green Revolution occurred in Europe in the 50 years after 1870 in response to rural labour shortages as countries industrialised. Agricultural production moved to more capital-intensive models based around the family farm in recognition of the central role peasant agriculture played in providing food to the urban working class (Harwood 2012). This shift not only occurred in Europe but also in the so-called 'settler' countries such as Australia and the US. In these countries, however, the limit to food production lay in the prevailing large-scale range grazing systems on prime agricultural land: sheep in Australia, and cattle in the US.

In Australia the 'squatters' had established themselves on massive swathes of prime farming land to graze sheep for wool export to the UK, while in the US cattle ranches locked up prime farming land for extensive cattle grazing. There was little or no regulation controlling the process as government reach to the respective hinterlands was relatively weak, with little more than a police/military presence. The gold rushes of Australia and the US in the 1850s brought these issues to a head with a loss of labour for animal herding, and an increasing urban population, putting pressure on food prices, and at times necessitating food imports. Both Australia and the US responded with a series of land reforms, which basically allowed neophyte farmers to settle a patch of land, and if they fenced and cleared it for cultivation within five years, they could receive a title for a nominal payment (Nairn 2011; Rosecrance 1960; Seabrook, McAlpine, and Fensham 2006). The increased agricultural production was also helped by: the cheaper transport of inputs and outputs due to the expansion of railways; an increased use of nitrogen and phosphorous fertilizers, due to cheaper production processes; and new technologies including steel ploughs and mechanical threshers. In Europe,

> Credit cooperatives supplied the working capital for the purchase of the new inputs and the enlargement in livestock numbers; marketing cooperatives created efficient trade channels for the new inputs; and the cooperative dairy factories brought the great advantages of the centrifugal cream separator within the reach of the small farmer.
> (Zanden 1991, p.237)

Over a period of 50 years from around 1870 these changes enabled the farm sector to feed the rapidly industrialising societies. These land tenure changes, together with higher labour costs, led to technical advances often supported by government incentives. These included advances such as the stump jump plough and combine harvester in Australia in the 1880s, which transformed agricultural production (Connor 2004; Hills 1906). The key element in all of these cases was the role of the state in driving these changes.

A rapid expansion of state sponsored agriculture research and extension services to introduce the new farming technologies onto peasant farms to modernise them was started in Germany to ease political pressure from rapid industrialisation and food riots over urban price rises (Harwood 2012). This expansion of agricultural research was also taken up by a number of other countries including the US and Australia, with land grant universities in the US and Mechanics Institutes in Australia (Boon 2009; Berger, DeLancey, and Mellencamp 1984; Candy 1993). The effect of this set of changes over a relatively short period of time was the emergence of the self-contained family farm, in which little external labour was used, except for certain times of the year, such as harvesting. This solved the problem of acute labour shortages that had plagued the sector through the nineteenth century. In the late nineteenth early twentieth centuries new high yielding or pest resistant seed varieties were developed to take advantage of the new technology to further increase productivity. For example, high yielding wheat varieties in Germany in the 1880s, rust resistant wheat in Australia in 1901, followed by high yielding hybrid wheat in Italy and maize in the US (Eagles, Cane, and Vallance 2009; Farrer 1902; Harwood 2012; Saraiva 2010; Sutch 2008). This first 'green revolution' doubled agricultural productivity in those countries with the strongest supporting institutions (Germany and the Low countries – the Netherlands and Belgium) over the 50 years to 1920. It also cemented the role of the family farm and family labour as central to farm production in the West, which continues to be the case in the 2010s.

These institutional processes of technological change supported by land tenure reforms and government supported extension services (e.g. see Nairn (2011)), and the associated increase in food production were to be repeated after the Second World War in parts of the developing world, such as China, Mexico, and India. Even though women did have a central role on the family farm throughout this period, they and their work were not recognised in either policy or practice, except in periods of dire need such as war time, when they acted as a reserve labour force. The women's Land Army was formed in the UK, US, and Australia during both the first and second world wars to take over farms left by men who went to war. When the war was over these women were effectively pushed back into the domestic sphere and their role on the farm hidden once more.

The issue of women as farmers is one that keeps emerging, but is systematically ignored in part due to the institutional structures that drive much policy and practice. These are arguably based around the broader patriarchal government and societal structures in which they are set, what Hearn (2015) refers to as 'transpatriarchies'. The next section will discuss how the Green Revolution emerged in developing countries and the factors that drove it.

The post-war 'Green Revolution'

The term 'Green Revolution' was coined by USAID Administrator William Gaud in 1968 almost as a 'marketing' phrase to gain US Congressional support for the foreign aid budget. It was a neologism to highlight the recent successes of US foreign aid that had led to in increased agricultural output in several of its recipient countries, but particularly in India over the previous two years. The International Maize and Wheat Improvement Center in Mexico (CYMMIT) in the 1950s, and the International Rice Research Institute (IRRI) in the Philippines in the 1960s, in particular, were important locations for the research that was based around breeding high yielding varieties of food grains, principally wheat and rice, and to a lesser extent maize (Cleaver 1972; Lerner 2018; Patel 2013). Over time the brand 'Green Revolution' took on the characteristics of a trope or metaphor for any agricultural development, particularly those development initiatives supported by the US government and US led research institutes in developing countries (Sumberg, Keeney, and Dempsey 2012).

Gaud's use of the term 'green' revolution also made a political point by contrasting it to the communist 'red' revolution that threatened to sweep through Asia and Latin America at the time (Cleaver 1972; Hill 2017; Lerner 2018; Lerner 1990; Pere and Shelton 2006). Supporting the Green Revolution became, for a period, a key part of the US Cold War struggle against, initially the Soviet Union, and later China to counter the rise of communism (Engel 2012; Hagen and Ruttan 1988; Wood 1986). The use of foreign aid as a Cold War weapon dated back to US President Truman and his 1949 point-four program that saw foreign aid as part of its foreign policy framework (Kilby 2017; Kilby 2004; Wood 1986;). As I argue in the next section there was an almost parallel 'green revolution' happening in China using similar processes but with home grown technology, which China was soon also transferring to developing countries.

In the early 1940s, leading philanthropists Nelson and John D. Rockefeller, who were also concerned with the communist threat, saw

US know-how and technology as the way forward in a geo-political contest where the assertion of US interests was the main game, while poverty alleviation was secondary (Rockefeller 1951). As Perkins (1990) noted:

> The Rockefeller Foundation programs in agriculture had a substantial influence on subsequent American foreign aid programs, and strategic thinking continues to dominate the design of these packages. Improving the well-being of people everywhere is not likely to flow easily from such efforts.
>
> (p.15)

While the Rockefellers' original plan in the 1940s did include sorghum and millet, crops grown by the poor, this did not happen as plant breeders and governments chose the higher profile wheat and rice grown by larger peasant farmers. The marginal, poorer, and often women farmers could not afford the inputs and management time required to achieve the higher yields for rice and wheat. It was not until the 1980s when the Green Revolution was in decline that attention was paid to sorghum and millet (Harwood 2012).

The Green Revolution in the Global South, funded mainly by the Rockefeller Foundation in the 1940s and later the US government up to the 1970s, was quite different to what happened in Europe and elsewhere in the late nineteenth century in important ways. The postwar Green Revolution took a more reductionist approach and tended to ignore local farming systems and local sources of knowledge, in favour of 'scientific' knowledge. This had the effect of limiting its success in many places as the suggested technologies were not suited in the same way across a wide range of farmers and their differing needs (Basu and Scholten 2012; DeWalt 1985; Spring 1986).

There was the modernist view at the time that the relevant technology to hand had the answers to food shortage, arguing that malnutrition was simply due to underproduction, and there was little place for local input of any sort, let alone that of women (Harwood 2012; Robertson 2016). For example, Edwin Wellhausen, a plant breeder with the Rockefeller Foundation, who toured Mexico in 1943 and collected 2,000 varieties of maize from indigenous communities for the Mexico Agriculture Program (MAP), seems to have had little interaction with these communities. From the available evidence and his own writings, at no stage did Wellhausen or his team talk to Mexican native Americans about how they might have bred these different varieties (Eddens 2017).

For Wellhausen it seemed that plant breeding was a Western concept almost by definition, and local knowledge had little if anything to offer, and so it was either downplayed or at best obliquely alluded to. The suggestion was that the improvements in maize production over thousands of years was more by accident than intent. In his own book on Mexican maize varieties Wellhausen notes that: 'Man brings these varieties or races together under conditions where cross fertilization is inevitable and a new hybrid race of formed' (Wellhausen et al. 1952, p.21). There was no suggestion that Mexican native Americans' agency or intent had any role in this process. Later research, however, has challenged this assertion noting how Mayan Indians in Chiapas over a period of around 20 years had taken high yielding maize varieties and from them and existing local varieties had bred new varieties more suited to their varying highland conditions, in much the same way they had been doing since pre-Hispanic times: 'Local maize varieties have been adapted to specific agro-ecological conditions for centuries' (Brush, Corrales, and Schmidt 1988, p.316).

This is only one of the early, among many, instances of local knowledge being actively ignored in the story of the Green Revolution: 'Arrogance and an inclination to dismiss local knowledge—whether that of scientists or farmers—had hampered more than a few programmes' (Harwood 2012, p.399). Local knowledge, contexts, and men's and women's common or different preferences continue to be ignored or downplayed. The extent to which imposed solutions affect the changes being sought, such as food security or improved nutrition varies across different contexts, but these externally sourced solutions can also marginalise those most vulnerable and risk averse. This has been one of the legacies of the Green Revolution of the past and threatens to be repeated in the future. However, as the Mayan example shows, local adaptation is still possible.

While Robertson (2016) argues that the Green Revolution was a continuation of Truman's 1949 Point Four program, the US government in the 1950s and 1960s rode more or less on the coattails of the Green Revolution that was led by the Rockefeller Foundation, which valued its autonomy from government. The Rockefeller Foundation, however, did have common cause with the US, and so its funding in both Mexico in the 1940s and 1950s, and later with some UN support in India in the 1950s and 1960s, was also to counter communist influence (Harwood 2012). It was this work that kick started the post-war Green Revolution from the West.

> Keeping Mexico friendly to the United States and preventing the possible "loss" of India to communism were the frameworks

within which the [Rockefeller] Foundation laid the groundwork for [its] development assistance.

(Perkins 1990, p.15)

By the 1960s in India, the Rockefeller support for the Green Revolution was increasingly at odds with a changing US Cold War foreign policy (Cleaver 1972; Perkins 1990). The US was trying to 'buy' Indian support for foreign policies through the provision of food aid while the Rockefeller Foundation was trying to wean India off food aid through increased domestic grain production.

The Rockefeller Foundation may have also played a role in China as it was also involved in crop improvement research in Nanjing, and in agriculture more generally, in the 1930s, based on the Tennessee Valley Authority experience. These experiences may have been revived in the 1950s by the People's Republic in the 1950s in its quest for increased food production (He et al. 2001; Ekbladh 2002). The next section will look at how the export of Green Revolution technology from the West and later from China, had a number of similarities but also some key differences.

The Green Revolution from the West

The 1950s and 1960s was a period of optimism by both the Soviet bloc and the West, with technology believed to be able to solve many of the world's social problems (Lerner 2018; Robertson 2016; Westad 2005). The Rockefellers took this a step further and believed technology could keep Soviet influence at bay: 'agricultural science had an important political role to play in the emerging US-USSR struggle' (Hill 2017, p. 11). It was also a period of US leadership of the West in the Cold War contest for the 'hearts and minds' of people of the Global South, but one that also offered opportunities for US capital in new and emerging developing country markets (Harwood 2012; Lavelle 2013; Pieterse 2012; Westad 2005).

The 1970 Nobel Peace Prize winner, US plant breeder Norman Borlaug and the only scientist to have won the prestigious award, became the poster boy for the 'success' of the Green Revolution and actively promoted it for the rest of his life (Lerner 2018; Patel 2013). He led the research on new wheat varieties in Mexico and personally led their introduction into India and Pakistan in the mid-1960s. He was also a passionate advocate for technical fixes and would not brook any criticism of the Green Revolution. He actively tried to thwart discussions of alternative more environmentally sustainable pathways for increased agricultural production by framing the narrative around the importance of agricultural technical advances as a way to discredit his critics (Sumberg, Keeney, and Dempsey 2012). However, Borlaug did not have it all his own way, in the 1960s and 1970s, as well as uncritical support, there was also scepticism of the benefits the Green Revolution had brought. For example, when the Paddock brothers' doomsday scenario in *Famine 1975* did not come to pass, they still argued that at best the Green Revolution delayed rather than denied their doomsday scenario. Their view was that any benefits were transitory at best, and due more to good luck than setting up long-lasting change (Harkins 1968; Paddock 1970; Paddock and Paddock 1967). This reflected the prevailing paternalistic view of the Global South, and a poor understanding of the complex interaction of population growth, development, and environmental degradation (Parayil 1998).

The activism of Borlaug and others, as well as the population debates, served to obscure the contextual and distributional issues resulting from the Green Revolution. The adverse social impacts of socio-economic policy on those who missed out, the poor and marginal and women farmers, were more often than not brushed aside. These impacts were noticed almost immediately (Cleaver 1972; DAC Secretariat 1970; Perkins 1990; Visvanathan 2003; Wellhausen 1976) but were ignored by the main funding bodies, the Rockefeller Foundation and the US government. They were far more concerned with the fears of rapid population growth, or the strategic links these projects could forge to combat communism. John D. Rockefeller had established the Population Council in 1952 to make the case for more focused population related measures in aid programs as well as in the work of the Rockefeller Foundation, rather than working on issues around sustainable development in the Global South (Cleaver 1972; Hoff 2010; Perkins 1990). Developing countries themselves were also less concerned with distributional issues and had policies to support larger farmers so faster productivity benefits could be realised at a national level (Lerner 2018; Patel 2013; Visvanathan 2003).

As part of the Cold War strategy to counter the advance of communism, the success of the Green Revolution is also debatable. Borlaug himself later praised, seemingly without any sense of irony, 'Red' China as the outstanding success story of the Green Revolution (Borlaug and Dowswell 2003). The Green Revolution from the West had little effect on China, without a major re-writing of history. India's political culture was also able make a noticeable shift to the left, due to the political space that opened from the Green Revolution success in increasing agricultural output, making it less politically dependent on the US and its foreign aid. It was also secure food supplies that cemented the Chinese Communist Party rule from the late 1960s and it was to be another 20 years before free market capitalism would make any inroads in either India or China.

The story of the Green Revolution is much more complex and varied. While there were significant advances in food production in most countries when crop yields more than doubled in the period 1950 to 1980 (Harwood 2012), paradoxically, at the turn of the twenty-first century, the number of people still living in hunger in developing countries has actually increased by more than 11 per cent if China's numbers are not included in the analysis (Rosset, Collins, and Lappé 2000). China's numbers can be validly left out due to the quite different agricultural policy settings in China vis à vis other countries that also had Green Revolution technical inputs, as we will see in the next section.

China's Green Revolution to the rest of the world

On the other side of the ideological fence of the Cold War, China's own 'green revolution' was decidedly 'redder' (Patel 2013, p.6). Following the establishment of the People's Republic of China (PRC) in 1949, the country invested heavily in agriculture productivity including plant breeding, as well as irrigation expansion, agricultural extension and land reform (Xu et al. 2016). The key ideological difference between China and the West was the role of the state in establishing and enforcing conducive policy settings in often quite authoritarian ways, which were neither possible nor desirable elsewhere.

China in the 1950s, however, was less focussed on exporting its nascent agricultural technology, despite the commitments to South–South technical cooperation it made at the Bandung Asia-Africa conference of 1955 (Kilby 2017; Lee, Wainwright, and Glassman 2017; Mawdsley 2012; Wright 1956). Agricultural aid from China did not get underway until the 1960s as a way to counter Taiwanese investment in the sector, particularly in Africa. In order to get African support for the China seat at the UN and weaken Taiwan diplomatically, China set up state owned sugar and tea plantations in Tanzania, Mali, Benin, Togo, Madagascar, Zanzibar and Sierra Leone. In 1971 after China's accession to the UN the PRC took over Taiwan's demonstration farms in Sierra Leone, and then built another six small and medium-sized rice demonstration stations across the country. These programs also included the provision of agricultural experts, so that by 1971 they were being sent to 18 countries. By the

1980s China had programs in 25 African countries (Bräutigam and Xiaoyang 2009, p.688), which then grew in size and scope, and from 2006 were more-or-less consolidated into an Agricultural Development Centre program, which is now across 23 countries of Africa (Xu et al. 2016, p.84; Brautigam 2015).

Overall these projects have not met their objectives. Like the World Bank and other donors with their integrated rural development projects at the time, only around half were successful (Bräutigam and Xiaoyang 2009, p.689). The failure has been arguably due to their inability to adapt to the local context, and a top down approach in which the aid recipients were being asked to accept the technologies on a 'take it or leave it' basis. It is therefore not surprising then that many communities did not take up the technologies, and when they seemed to be imposed this led to tensions and conflict, either with the host government or China as the donor (Brautigam 2015; Bräutigam and Xiaoyang 2009; Xu et al. 2016). For example, in Cameroon the government promised a Chinese company 10,000 hectares if it could rehabilitate earlier Taiwanese rice production projects to operate on a commercial scale. The Chinese company ran into problems with the local community who found the working conditions difficult and claimed customary title to some of the land. The result was that after four years the program had not expanded beyond the initial 100 hectares that was granted to the project, as the Cameroon government did not want to enter a dispute with the local community over customary title and access. This mirrored earlier poor experiences with US corporate investors with oil palm in Cameroon (Brautigam 2015). This type of experience is a common story with many donors, and is a reason that the adoption of Green Revolution technologies has been patchy across much of Africa (e.g see Allen, Porter and Thompson 1991/2014).

In the 2010s Bräutigam and Xiaoyang (2009) observed 'three overlapping patterns in this agricultural engagement: diplomacy-based aid, public–private partnerships and overseas investment by Chinese companies' (p.687). The first of these three patterns reflect foreign aid in the traditional sense while the other two are more commercially based. At the diplomatic level China's aid aimed to

> embed their expertise in agriculture as a component of diplomatic efforts to shape their engagement with Africa and provided for a large number of African experts training in China and a similar number of China's experts working in Africa.
>
> (p.695)

This traditional approach towards aid as South–South technical cooperation was a major focus of the Bandung agreement, and the subsequent eight principles of foreign aid enunciated by Chinese Premier Zhou Enlai in 1964 (Brazinsky 2017; Enlai 1964; Kilby 2017).[1] These principles were aimed to be in contrast to Western technical assistance, which used highly paid consultants and external driven inputs. China's aid has generally followed the eight principles, and these also have been reiterated in the 2010s (Government of China 2011; Mattlin and Nojonen 2015; Varrall 2016). As noted above, however, the weakness was similar to that of Western aid, notwithstanding the eight principles. That is, the imposition of China's technologies and techniques are difficult to absorb in the various cultural contexts of Africa (Allen, Porter and Thompson 2014).

China's agricultural support program, however, has remained relatively low key compared to China's infrastructure investments in Africa, particularly those under the most recent Belt Road Initiative. Thus, the contest of China with the West in the Green Revolution over the past 50 years has been less obvious (Xu et al. 2016). Like the Belt Road Initiative in the 2010s, in the 1950s and 1960s there was also a set of infrastructure activities that complemented the Green Revolution both from the West and the Soviet bloc, including the building of dams and irrigation systems across the Global South.

Dams and irrigation

Part of the enabling environment for the Green Revolution was the massive investment through the 1950s and 1960s in large dams for both energy and irrigation. Infrastructure and public utilities were preferred in foreign aid programs, over welfare related areas such as health and education, despite the obvious but less direct impact on productivity that education and health improvements bring (Engel 2012). Dams and irrigation were part of the modernising boom of the 1950s and supported by both Cold War rivals, the US and the Soviet Union. The World Bank in the 1950s under Eugene Black, who was to be its longest serving President, was very enthusiastic to the point that one-third of the Bank's loans went to large dams. This approach by the Bank was based on similar experiences in the US in the 1930s with the Tennessee Valley Authority (TVA), and the prevailing technocratic approach to agricultural and industrial development (Ekbladh 2002; Kline and Moretti 2013; Robertson 2016).

It was the perceived success of the TVA that made it an important example for international development work to follow. Together with the Green Revolution plant breeding technologies of Borlaug and others, large scale irrigation and power generation schemes were seen as, not only a way to achieve positive 'grass roots' development, but an important weapon in the Cold War.

> TVA inspired programs highlighted the vital difference between U.S. and communist development styles. The Soviet model promised rapid economic growth, but with the penalty of authoritarian regimentation.
>
> (Ekbladh 2002, p.350)

Dams and TVA type investments were also used to ease international tension. For example, the conflict between Pakistan and India over the access to the Indus river waters led to 'extending World Bank technical assistance to resolve the Indus water resources question "on an engineering basis"' (Ekbladh 2002, p.353; Qureshi 2016). This program at best ameliorated the situation rather than solved it. For example, proposed dams by China in Pakistan in the 2010s on the upper reaches of the Indus river are further stoking these earlier tensions (Roic, Garrick, and Qadir 2017; Wolf 2016).

Dams provided the electricity for industrial expansion and the water for the agricultural intensification driven by Green Revolution high yielding varieties, to 'make the world safe for capitalism' (Westad 2005, p.31). It was the irrigation projects in Mexico and India that enabled the HYVs to be successful, however, many of the large dams were environmental disasters, and failed to live up to the social and economic expectations (Caufield 1996; Robertson 2016). While the TVA was meant to be based on 'democratic participation' (Ekbladh 2002, p.337), its implementation in practice was very much top down, and there was little community voice in either its design or operations (p. 345). Such a top down approach has been common to most large-scale irrigation projects regardless of the donor. Despite the rhetorical contrast from the West to the Soviet methods, they were actually not that far apart in their approach when it came to the practice of building large dams and irrigation infrastructure (p.351).

Dams, however, made a strong bargaining chip in the Cold War between the Soviet Union and the West (Westad 2005; White 1974). The Aswan High Dam in Egypt was one site for such a battle, with the Soviet Union eventually winning that contract and the associated propaganda battle, after protracted negotiations through the early 1950s (Brazinsky 2017; Goldman 1967; Nef 2004; Westad 2005; White 1974). The Aswan High Dam was relatively successful as an irrigation and electricity project, increasing Egypt's irrigated area one third, but also had some adverse environmental impacts such as waterlogging, coastline erosion, and some increased salinity, most of which, fortunately, have been ameliorated (Robinson et al. 2008; White 1988). Despite these emerging risks and the fading image of the TVA, in the 1970s irrigation and dams were still important in the World Bank portfolio with over third of agricultural sector lending still being for irrigation (Ekbladh 2002; Hotes 1983).

Many, if not most, of these dams did not meet their output targets nor provide the expected economic returns to ensure their economic viability. But it was the promise of economic viability, which drove World Bank funding decisions, so a failure to meet the expected output saddled the borrowing country with using precious foreign exchange reserves to payback the World Bank, and other donor loans, with little revenue to build these reserves up again. For example, in Thailand the record of dams over thirty years was that irrigation projects provided water to less than half their command area (with the best being 69 per cent), and only two-thirds of planned electricity output being provided with many programs providing less than half (Rich 2014, p. 12). This puts a strain on the rest of the economy to repay the debt.

In another case, the Helmand Valley Dam project in Afghanistan:

> The very first dam lifted the water table almost to surface level, ringing the reservoir with salt. Yet construction continued and even expanded, spreading the waterlogging and salination. The dams also blocked the silt that once rejuvenated downstream fields. Astonishingly, over the course of the project crop yields actually declined.
>
> (Robertson 2016, p.431)

These massive dam investments took on an almost religious zeal spurred by Cold War rivalry and were seen as important for the prestige of the recipient country (Caufield 1996), who 'willingly mortgaged their future in order to secure [their] short-term survival' (Westad 2005, p.157).

In addition to not meeting their output targets and increasing debt, these dams also sapped local budgets in order to provide counterpart costs (Caufield 1996; Griffin and Enos 1970). For example, 50 per cent of Colombia's national budget in the late 1950s went to cover counterpart costs mainly for World Bank dams, and only a few per cent went to education (Caufield 1996, p.64). These institutional issues around infrastructure and the available budget for other types of agricultural support services support influenced the success or otherwise of the Green Revolution in those areas.

For the Green Revolution from the West, investment in software (support to farmers) came second to investment in hardware, which was quite different to the experiences in the late nineteenth and early twentieth centuries. This 'love affair' in developing countries with technology both from the East and West was in part what drove the technocratic approach of the Green Revolution together with the largesse of donors such as the Rockefellers and Foundations that were able to sell their ideas to the US government and others such as the World Bank in the 1950s and 1960s. The story of Mexico, India and China are cases in point and will be looked at in more detail in Chapter 3.

Note

1 The principles are: i) mutual benefit rather than charity; ii) no conditions or privileges for China; iii) interest-free or low-interest loans with the possibility of extending the repayment period if necessary; iv) encouraging independence of the recipient's economy; v) low capital input with quick rates of return; vi) free replacement of unsuitable equipment; vii) 'on-the-spot' training of local counterparts; viii) willingness of Chinese experts to accept local living standards.

References

Allen, Bryant, Doug Porter, and Gaye Thompson. 1991/2011. *Development in Practice: Paved with Good Intentions*. Routledge Revivals, Abingdon: Routledge.

Basu, Pratyusha, and Bruce A. Scholten. 2012. "Crop–livestock systems in rural development: Linking India's Green and White Revolutions." *International Journal of Agricultural Sustainability* 10(2): 175–191.

Berger, Marguerite, Virginia DeLancey, and Amy Mellencamp. 1984. *Bridging the Gender Gap in Agricultural Extension*. Washington, DC: International Center for Research on Women.

Boon, Katherine Fay. 2009. "A critical history of change in agricultural extension and considerations for future policies and programs", PhD Thesis, School of Agriculture Food and Wine, Adelaide: University of Adelaide.

Borlaug, Norman E., and Christopher R. Dowswell. 2003. "Feeding a world of ten billion people: A 21st century challenge." In Phillips, R.L., R. Tuberosa, and M. Gale (eds) *Proceedings of the International Congress: In the Wake of the Double Helix: From the Green Revolution to the Gene Revolution, 27–31 May*, pp. 3–23. Bologna: Avenue Media.

Brautigam, Deborah. 2015. *Will Africa Feed China?* Oxford: Oxford University Press.

Bräutigam, Deborah A., and Tang Xiaoyang. 2009. "China's engagement in African agriculture: 'Down to the countryside'." *The China Quarterly* 199: 686–706.
Brazinsky, Gregg A. 2017. *Winning the Third World: Sino-American Rivalry during the Cold War.* UNC Press Books.
Brush, Stephen B., Mauricio Bellon Corrales, and Ella Schmidt. 1988. "Agricultural development and maize diversity in Mexico." *Human Ecology* 16(3): 307–328.
Candy, Philip C. 1993. "The problem of currency: Information literacy in the context of Australia as a learning society." *The Australian Library Journal* 42(4): 278–299.
Caufield, Catherine. 1996. *Masters of Illusion: The World Bank and the Poverty of Nations.* London: Macmillan.
Cleaver, Harry M. 1972. "The contradictions of the Green Revolution." *The American Economic Review* 62(1/2): 177–186.
Connor, David J. 2004. "Designing cropping systems for efficient use of limited water in southern Australia." *European Journal of Agronomy* 21(4): 419–431.
DAC Secretariat. 1970. DAC Chairman's Report 14 August, in *OECD Archives (DAC 70-37)*, edited by OECD. Paris.
DeWalt, Billie. 1985. "Farming systems research: Anthropology, sociology, and farming systems research." *Human Organization* 44(2): 106–114.
Eagles, Howard A., Karen Cane, and Neil Vallance. 2009. "The flow of alleles of important photoperiod and vernalisation genes through Australian wheat." *Crop and Pasture Science* 60(7): 646–657.
Eddens, Aaron. 2017. "White science and indigenous maize: The racial logics of the Green Revolution." *The Journal of Peasant Studies*: 1–20.
Ekbladh, David. 2002. "'Mr. TVA': Grass-roots development, David Lilienthal, and the rise and fall of the Tennessee Valley Authority as a symbol for US overseas development, 1933–1973." *Diplomatic History* 26(3): 335–374.
Engel, Susan. 2012. *The World Bank and the Post-Washington Consensus in Vietnam and Indonesia: Inheritance of Loss.* London: Routledge.
Enlai, Zhou. 1964. The Chinese Government's Eight Principles for Economic Aid and Technical Assistance to Other Countries January 15, in the *History and Public Policy Program Digital Archive, Selected Diplomatic Papers of Zhou Enlai 1990*, edited by Wilson Center Digital Archive. Wilson Center: Wilson Center Digital Archive. http://digitalarchive.wilsoncenter.org/document/121560.
Farrer, William. 1902. "Federation variety of wheat." *Agricultural Gazette of NSW* 12: 977.
Goldman, Marshall I. 1967. *Soviet Foreign Aid.* New York: Praeger.
Government of China. 2011. *China's Foreign Aid* (White Paper), April. Beijing: China's Information Office of the State Council, People's Republic of China.
Griffin, Keith B., and John L. Enos. 1970. "Foreign assistance: objectives and consequences." *Economic Development and Cultural Change* 18(3): 313–327.
Hagen, James M., and Vernon W. Ruttan. 1988. "Development policy under Eisenhower and Kennedy." *The Journal of Developing Areas* 23(1): 1–30.

Harkins, Robert W. 1968. "Famine-1975! America's decision: Who will survive?" *Journal of the American Medical Association* 203(2): 157–157.

Harwood, Jonathan. 2012. *Europe's Green Revolution and Others Since: The Rise and Fall of Peasant-Friendly Plant Breeding*. Abingdon: Routledge.

He, Z.H., S. Rajaram, Z.Y. Xin, and G.Z. Huang, eds. 2001. *A History of Wheat Breeding in China*. El Batan, Mexico: CIMMYT.

Hearn, Jeff. 2015. *Men of the World: Genders, Globalizations, Transnational Times*. London: Sage.

Hill, Brent. 2017. "The Green Revolution and the Rockefellers." *Isocraccy*, Feb. 28.

Hills, E.H. 1906. "The geography of international frontiers." *The Geographical Journal* 28(2): 145–155.

Hoff, D.S. 2010. "'Kick that population commission in the ass': The Nixon Administration, the Commission on Population Growth and the American Future, and the defusing of the population bomb." *Journal of Policy History* 22(1): 23–63.

Hotes, Frederick L. 1983. "World bank irrigation experience." *International Journal of Water Resources Development* 1(1): 65–75.

Kilby, Patrick. 2017. *China and the United States as Aid Donors: Past and Future Trajectories, Policy Studies 77*. Honolulu: East West Center.

Kline, Patrick, and Enrico Moretti. 2013. "Local economic development, agglomeration economies, and the big push: 100 years of evidence from the Tennessee Valley Authority." *The Quarterly Journal of Economics* 129(1): 275–331.

Lavelle, Kathryn C. 2013. "American Politics, the Presidency of the World Bank, and Development Policy." In *Policy Research Working Paper WPS6377*. Washington: World Bank Group Archives, Information Management & Technology Department.

Lee, Seung-Ook, Joel Wainwright, and Jim Glassman. 2017. "Geopolitical economy and the production of territory: The case of US–China geopolitical-economic competition in Asia." *Environment and Planning A: Economy and Space* 50(2): 416–436.

Lerner, Adam B. 2018. "Political neo-Malthusianism and the progression of India's Green Revolution." *Journal of Contemporary Asia* 48(3): 1–23.

Mattlin, Mikael, and Matti Nojonen. 2015. "Conditionality and Path Dependence in Chinese lending." *Journal of Contemporary China* 24(94): 701–720.

Mawdsley, Emma. 2012. "The changing geographies of foreign aid and development cooperation: Contributions from gift theory." *Transactions of the Institute of British Geographers* 37(2): 256–272.

Nairn, Bede. 2011. "Robertson, Sir John (1816–1891)." *Australian Dictionary of Biography*, National Centre of Biography, Australian National University, http://adb.anu.edu.au/biography/robertson-sir-john-4490/text7337, accessed October 22, 2018.

Nef, Jorge. 2004. "International Development Studies and ethical dilemmas in academia." *Canadian Journal of Development Studies* 25(1): 81–100.

Paddock, William C. 1970. "How green is the green revolution?" *BioScience* 20(16): 897–902.

Paddock, William, and Paul Paddock. 1967. *Famine-1975! America's Decision: Who Will Survive?* Boston: Little, Brown and Co.

Parayil, Govindan. 1998. "Sustainable Development: The fallacy of a normatively-neutral development paradigm." *Journal of Applied Philosophy* 15(2): 179–194.

Patel, Raj. 2013. "The long green revolution." *The Journal of Peasant Studies* 40(1): 1–63.

Pere, Garth le, and Garth Shelton. 2006. "Afro-Chinese relations: An evolving South–South partnership." *South African Journal of International Affairs* 13(1): 33–53.

Perkins, John H. 1990. "The Rockefeller Foundation and the green revolution, 1941–1956." *Agriculture and Human Values* 7(3–4): 6–18.

Pieterse, Nederveen. 2012. "Twenty-first century globalization: A new development era." *Forum for Development Studies* 39(3): 367–385.

Qureshi, Waseem Ahmad. 2016. "The Indus Waters Treaty and the role of World Bank as mediator." *Willamette Journal of International Law & Disaster Research* 24: 211.

Rich, Bruce. 2014. *Mortgaging the Earth: World Bank, Environmental Impoverishment and the Crisis of Development*. Abingdon: Routledge.

Robertson, Thomas. 2016. "Cold War landscapes: Towards an environmental history of US development programmes in the 1950s and 1960s." *Cold War History* 16(4): 417–441.

Robinson, Sherman, Ken Strzepek, Moataz El-Said, and Hans Lofgren. 2008. "The high dam at Aswan." *Indirect Impact of Dams: Case Studies from India, Egypt, and Brazil*. Washington, DC, and New Delhi, India: World Bank and Academic Foundation: 227–273.

Rockefeller, Nelson A. 1951. "Widening boundaries of national interest." *Foreign Affairs* 29(4): 523–538.

Roic, Kristina, Dustin Garrick, and Manzoor Qadir. 2017. "The ebb and flow of water conflicts: A case study of India and Pakistan." In *Imagining Industan*, 49–66. Springer.

Rosecrance, Richard. 1960. "The radical tradition in Australia: An interpretation." *The Review of Politics* 22(1): 115–132.

Rosset, Peter, Joseph Collins, and Frances Moore Lappé. 2000. "Lessons from the Green Revolution." *Third World Resurgence*: 11–14.

Saraiva, Tiago. 2010. "Fascist labscapes: geneticists, wheat, and the landscapes of fascism in Italy and Portugal." *Historical Studies in the Natural Sciences* 40(4): 457–498.

Seabrook, Leonie, Clive McAlpine, and Rod Fensham. 2006. "Cattle, crops and clearing: regional drivers of landscape change in the Brigalow Belt, Queensland, Australia, 1840–2004." *Landscape and Urban Planning* 78(4): 373–385.

Spring, Anita. 1986. *Trials and Errors: Using Farming Systems Research to Reach Farmers who are often Neglected*. Edited by Cornelia Butler Flora and Martha Tomecek. Vol. Farming systems research paper series, paper no. 9, *Farming Systems Research Symposium on Farming Systems Research and Extension: Implementation and Monitoring* (7–10 October 1984). Manhattan, KS: USAID.

Sumberg, James, Dennis Keeney, and Benedict Dempsey. 2012. "Public agronomy: Norman Borlaug as 'brand hero' for the Green Revolution." *The Journal of Development Studies* 48(11): 1587–1600.

Sutch, Richard C. 2008. *Henry Agard Wallace, the Iowa Corn Yield Tests, and the Adoption of Hybrid Corn*. Washington: National Bureau of Economic Research.

Varrall, Merriden. 2016. "Domestic actors and agendas in Chinese aid policy." *The Pacific Review* 29(1): 21–44

Visvanathan, Shiv. 2003. "From the Green Revolution to the Evergreen Revolution: Studies in Discourse Analysis." IDS Seminar on Agriculture Biotechnology and the Developing World. October, Brighton, IDS.

Wellhausen, Edwin J. 1976. "The agriculture of Mexico." *Scientific American* 235(3): 128–153.

Wellhausen, Edwin John, Lewis M. Roberts, X. Hernandez, and Paul C. Mangelsdorf. 1952. *Races of Maize in Mexico. Their Origin, Characteristics and Distribution*. Boston: Bussey Institution, Harvard University.

Westad, Odd Arne. 2005. *The Global Cold War: Third World Interventions and the Making of our Times*. Cambridge: Cambridge University Press.

White, Gilbert F. 1988. "The environmental effects of the high dam at Aswan." *Environment: Science and Policy for Sustainable Development* 30(7): 4–40.

White, John. 1974. *The Politics of Foreign Aid*. London; Toronto: Bodley Head.

Wolf, Siegfried O. 2016. "The China-Pakistan economic corridor: An assessment of its feasibility and impact on regional cooperation." In *SADF Working Paper No. 1*. Brussels: South Asia Democratic Forum.

Wood, Robert Everett. 1986. *From Marshall Plan to Debt Crisis: Foreign Aid and Development Choices in the World Economy*. Vol. 355. Berkeley and Los Angeles: University of California Press.

Wright, Richard. 1956. *The Color Curtain: A Report on the Bandung Conference*. Cleveland, OH: The World Publishing Company.

Wu, Huifang, and Jingzhong Ye. 2016. "Hollow lives: Women left behind in rural China." *Journal of Agrarian Change* 16(1): 50–69.

XuXiuli, Xiaoyun Li, Gubo Qi, Lixia Tang, Langton Mukwereza. 2016. "Science, technology, and the politics of knowledge: The case of China's agricultural technology demonstration centers in Africa." *World Development* 81: 82–91.

Zanden, J.L. van. 1991. "The first green revolution: The growth of production and productivity in European agriculture, 1870–1914." *The Economic History Review* 44(2): 215–239.

3 Three Green Revolution case studies

Introduction

To understand the Green Revolution, it is important to see how it emerged in different contexts. This chapter will look at three case studies in quite different contexts: Mexico, the home of the Green Revolution, and India and China two emerging powers which were keen to make a mark on the world stage to challenge what they saw as Western hegemony. What emerges in each of these three case studies is the role of the state in ensuring that technical advances had supporting infrastructure, and the focus in Mexico and India in particular on the larger peasant farmers, leaving out, at least initially, small-scale marginal peasant farmers and women farmers all together.

Mexico

Mexico has been called the home of the Green Revolution, having had a fourfold increase in agricultural output in the 25 years to 1965 due to the adoption of HYVs (high yielding varieties), particularly wheat. However, it went through a number of stages in its agricultural transformation, some of which were much earlier, which laid the groundwork for the remarkable changes from 1940 (Cullather 2010; Sonnenfeld 1992;). The key stage was the land reform that started in 1917 but accelerated under the populist president General Cardenas, which led to an increase of 2.5 million acres of land coming under cultivation (Cullather 2010, p.31). The second part of the strategy was 'extending the agricultural frontier through irrigation projects and the diffusion of biochemical (land-saving) technologies' (p.31). The amount of land under irrigation increased nearly 50 per cent, from 28 million acres in 1930 to 41 million acres in 1960 (p.32). This made the country ready to adopt the fertilizer responsive HYV seeds developed by the Mexican Agricultural Program supported by the Rockefeller Foundation.

The Rockefeller Foundation had been working in the agricultural sector from the early 1900s and began early talks in Mexico from 1933 but suffered a setback in 1934 with the seizure of Nelson Rockefeller's oils assets in 1934 by President Cardenas (Perkins 1990). It was US vice-president Wallace's road trip to Mexico for the inauguration of the more US friendly President Camacho, the follow up visit by Nelson Rockefeller and the survey team that gave the impetus to the Rockefeller Foundation's work in Mexico. The recommendation from this survey was 'a four-man commission in or near Mexico City to advise the Mexican Department of Agriculture' (p.8). The Commission led by George Harrar commenced its work in 1943 as a semi-autonomous Office of Special Studies within the Mexican Ministry of Agriculture. Norman Borlaug joined the Commission in 1944 (Cullather 2010; DeWalt 1985; Eddens 2017; Paarlberg 1970; USAID 1982). The Office of Special Studies then established the Mexico Agricultural Program (MAP) and a research station with a team made up exclusively of scientists, such that

> Distribution of yield and the social complexities that governed distribution were simply not of interest to these scientists...[but] Foundation officers and trustees were deeply committed to geopolitical concerns of political and military significance.
> (Perkins 1990, p.15)

While the MAP may not have engaged social scientists, Norman Borlaug was quite hands on and tested his research via farmer demonstration days that grew quickly in popularity to reach thousands of farmers within a few years (Paarlberg 1970, p.9). Over the 1940s and 1950s the MAP produced a number of new varieties of wheat and maize that were well suited to the newly opened irrigated farmlands in the Northwest of Mexico, of which wheat was the most successful (Cleaver 1972, p.181). The first high yielding varieties of wheat were released in the late 1940s and rust resistant wheat varieties through the 1950s (Phillips 2013), so that by 1958 rust resistant strains made up 70 per cent of wheat grown in Mexico (Paarlberg 1970, p.10).

Semi-dwarf varieties of wheat were made available to farmers a little later in 1962 as another step in what Borlaug called the 'the "quiet" wheat revolution' which started in the 1940s (Borlaug 2002, p.4). Here Borlaug took dwarf varieties of wheat from Japan and crossed them with local Mexican varieties and he was able release them after only eight years of breeding trials (Paarlberg 1970, p.12). Semi-dwarf varieties of HYV wheat were favoured as they converted nitrogen to grain more efficiently, and they did not 'lodge' or fall over on maturity and so being less difficult to harvest (Sumberg, Keeney, and Dempsey 2012; USAID 1982). The cumulative effect of these changes to wheat cultivation was a near trebling of average yield across Mexico to 30 bushels per acre, and in the favourable areas as high as 100 bushels per acre (Paarlberg 1970, p.13).

These successes were confined to the larger mainly wheat farmers in wealthier regions of northern Mexico (Cleaver 1972; Cullather 2010; Tuckman 1976). The approach taken was to provide a technological package of extension services, HYV seeds, fertilizers, pesticides, and even crop insurance, all of which the peasant farmer had to fund through cash flow or credit, an approach that favoured the larger farmer who could afford this (Clawson and Hoy 1979; Sonnenfeld 1992). While the land reform of the 1930s gave more land to the small-scale peasant farmer, the *ejidos*, in a dual agricultural policy of supporting both large commercial farmers and the small-scale peasant farmer, in practice the *ejido* became increasingly marginalised as they did not have the capital resources to fully exploit Green Revolution technologies. This was particularly the case with maize farmers who had a lower rate of HYV uptake compared to wheat farmers. For example, by 2010 only 31 per cent of maize plantings was devoted to HYVs (Becerril and Abdulai 2010, p.1024). In addition to the issue of capital availability, the benefits of HYVs of maize, while real, had a wide variation depending on climate and other exogenous factors, so there needed to be a more localised focus on improved varieties for particular agro-climatic zones.

Agriculture in the rest of Mexico also did not benefit to the same extent as the North because of lack of supporting infrastructure from government (Hill 2017).

> Functional dualism has meant sustained profits for agricultural and industrial capitalists and increased poverty and social dislocation for smallholders and agricultural workers.
> (Sonnenfeld 1992, p.35)

The effect was a shift away from HYVs for the *ejidos*, as policy and programs were not targeted to the local peasant contexts, as Clawson and Hoy found with their study of the Nealtican project area. In that project, a package of HYV maize was introduced to farmers, which was unsuitable due to reasons including: local micro-climate; soils not suited to nitrogen-based fertilizer; and compulsory crop insurance being required, but covering only one quarter of their crop. In addition, the maize introduced had poorer characteristics than their traditional crops and could not be adapted to their multiple crop planting practices to manage rainfall risk in rain fed areas (Brush, Corrales, and Schmidt 1988; Clawson and Hoy 1979). Despite this set of issues there was no flexibility and farmers were required to adopt the whole package on a take it or leave it basis. The result was poor uptake of the HYVs in the project area.

Major effects on the *ejidos* also occurred when the wheat started to replace corn and beans as the local staple, which was also price controlled under a cheap food policy for urban consumers, thus producing lower returns to the farmers.

> The production of basic food crops by peasants through rain-fed agriculture on *ejidal* and private plots was neglected by government policy makers in the 1970s. Not only were government agricultural loans less available to smallholders, but price supports for basic food grains were kept low in favor of urban workers to the detriment of rural producers, especially smallholders. The rain-fed, peasant agricultural sector, which had been one of the sources of expansion of agricultural production in the 1940s, 1950s, and 1960s, and a significant contributor to Mexico's food self-sufficiency, faced crisis.
>
> (Sonnenfeld 1992, pp.37, 38)

By the 1980s there were moves to reverse this trend away from supporting the poorer smallholder maize farmers, and programs were introduced targeting them, and maize farmers were able to take advantage of subsidies and other support that become available. These poorer farmers, however, were more risk averse and lived in a range of agro-environmental zones. They tended to plant a range of maize varieties to manage risks such as drought and the high winds that can lead to crops lodging. This comprised a mix of the original landrace varieties, intermediate varieties where local varieties had been crossed with HYVs, and HYVs. This strategy spread the risk, to give the farmer both drought resistance and resistance to lodging, depending on the weather during the growing season (Becerril and Abdulai 2010; Brush, Corrales, and Schmidt 1988; Smale, Bellon, and Aguirre Gomez 2001).

The story of the Green Revolution in Mexico is one of two crops in two regions and the importance of not only supporting government policies, but also the need for flexibility and adaptability to a range of agro-environmental zones. While wheat is very successful in controlled irrigated regions, maize, the traditional crop, requires more farm input for what the best HYV for their area is, as well as supportive government policies, if the poorer farmers are to get the full benefit from HYVs. This requires quite different sets of government policies.

India

In the case of India, the Green Revolution was a policy response to what was widely seen as an existential crisis in the mid-1960s, when agricultural output was unable to meet the demand of a growing population, and the capacity to import food on the scale required was stretched. By way of background, in the 1950s under Prime Minister Jawaharlal Nehru, India's national policy priority was around state led industrial development as a nation building strategy, in part to maintain India's strong non-aligned status (Abraham 2008; Berger 2004; Cullather 2010; Cullather 2017). This had implications for agricultural development as scarce foreign exchange reserves were being used for importing capital equipment for heavy industry rather than fertilizer and other agricultural inputs to increase agricultural productivity (Lerner 2018).

The focus for agriculture through the 1950s was more for community led development and extension services. This included expanding cooperatives and agricultural extension programs, together with land reform, rather than increasing the productivity of the land more directly through improved technology such as HYVs of the key cereal crops (Cullather 2010; Kaviraj 1988; Lerner 2018; Perkins 1990;). The nascent land reform movement, however, faltered (as with neighbouring Pakistan) due to recalcitrant Indian states, under pressure from landlords and feudal owners, not passing the necessary laws and regulations (Das 1999; Niazi 2004), that is, as Frankel (1971/2015) put it taking 'a go-slow to land reform' (p.4).

The community development program, while successful as a pilot program in the early 1950s, could not scale up fast enough to cover the whole country, nor did it lead to the necessary productivity increases required to stave off the regular food shortfalls that India faced at the time (Perkins 1990). As the flagship agricultural initiative, thousands of village extension workers were sent to work with local farmers. The pilot, with resettled Hindu displaced people from Partition, was lauded by Nehru in 1950 as a way forward to increase food production. It was a program that could be expanded 'without additional expenditure' from the government and was supported by both the Ford Foundation and USAID, as part of the Eisenhower government's strategy to bring India into the US geo-political fold (Loveridge 2017; Lerner 2018, p.7). The original idea of the program was to focus on those areas with the best natural resources. Under political pressure, however, this had to be abandoned, to be expanded to all India as a more extensive program, rather than the original more intensive approach to be rolled out in stages (Frankel 1971/2015, p.3). The focus of the program, which was also to be its weakness, was less about inputs and more on an increased use of labour through expanded peasant-based cultivation (p.4). Despite Nehru's question in his 1950 visit to the program about the lack of women in the program, his observation fell on deaf ears and the program, while well intentioned like others at the time and before, failed to recognise the role of women as farmers and focused solely on men with its extension and training services and other support. In all 60 million (male) peasants across India were mobilised through the program.

However, despite its extensive reach, the community development program was slow to show good results, due probably to its broad reach and the focus on labour rather than a broader package of technologies that supported a range of farming systems. What is interesting, however, is that where the community development program was strongest was also where the Green Revolution was most successful. The community development program built the communities' knowledge and understanding to be receptive to technological change (Tan and Kudaisya 2008) to the point that it was seen as 'influencing the course of the Green Revolution' (Loveridge 2017, p.58). It was Chidambaram Subramaniam, former Finance Minister who was appointed as Minister for Food and Agriculture in 1964 to solve the food crisis, who made the link between science (in this case plant science) and extension: '[India]...had to break the old ideologies of science pursued for its own sake and weld science and extension work' (Visvanathan 2003, p.6). What a focus on women farmers as well at the time would have achieved is anyone's guess.

A key obstacle was the feudal zamindar system of absent landlords in states mainly in Eastern India that held sway, and that was where the Green Revolution failed. It was not in the landlord's interest to engage in more productive technologies as their income came from the rent they charged and not from the productivity of the land, leading to what Das (1999, p.174) referred to as a rent usury barrier. Thus the successes of the Green Revolution were largely confined to areas where there were landowning peasants and from which, with their new-found wealth, they were able to build their political power (Das 1999; Loveridge 2017). It was only in a few cases that the landlords saw real advantages, rather than a threat, in their serfs or tenants adopting new technology (Pearse 1980/2015). The other issue was the inadequate analysis of the sociocultural context of much of India by Rockefeller Foundation staff, who were leading the research. They tended to be focused on population issues rather than the social factors that drove these issues.

Moreover, the population analysis brushed aside any consideration that the Indian people might be acting intelligently, even if they were poor, illiterate, and desiring a large family. Undoubtedly, the most important flaw in the assessment of India by Foundation officials was the failure to realize that distribution of food was as important as production.

(Perkins 1990, pp.13, 14)

This analysis was not entirely the fault of the Rockefeller Foundation, as Indian government policy was to focus on fewer larger successful farmers in places where success with HYVs was more assured. The backdrop to this change in agricultural policy to HYVs and technology was the delicate political situation India found itself in. While Prime Minster Nehru wanted to be a global leader of a non-aligned Global South, the Indian economy was too fragile for India to be truly non-aligned, so it depended almost in equal measure on Soviet and US support. By the early 1960s, after a decade of low growth in crop and grain production, which had been unable to keep pace with population growth, and slow and faltering progress in its own land reforms, India was heading inexorably to an agricultural crisis (Evenson, Pray, and Rosegrant 1998; Paddock and Paddock 1967).

At a geopolitical level this presented an opportunity for the US, which had been using foreign aid and, in particular, food aid, as an inducement for countries to move into its geo-political orbit and away from the non-aligned movement, which was seen to be too close to Soviet influence. The US large investment in the Community Development Program of the early 1950s was in part driven by an anti-communist agenda (Perkins 1990). However, by the early 1960s India, while still formally non-aligned, was more in the Soviet 'camp', certainly at a trade and diplomatic level, and this was an ongoing frustration to the US government. The Soviet Union's 'hands-off' policies in terms of dictating India's domestic policies played a big role in the shape of this relationship, as did a generous aid program in support of India's industrialisation, with loan terms more generous than those offered by the US (Berger 2004).

By the mid-1960s the Johnson administration in the US became increasingly disillusioned with its efforts of the 1950s, and the lack of a diplomatic or strategic return. Lyndon Johnson was desperate to have India in the US camp and for it to adopt a free market and open trade policies friendly to US investment, which the Nehru government had hitherto resisted. With India's high levels of grain imports and US food aid being part of that, Johnson had the opportunity to put India on a 'short-tether' as he called it, in the US response to India's drought of 1965. The rains failed and there was a 20 per cent drop in grain production driving food prices up, threatening famine in some states such as Bihar, which was hit very hard (Lerner 2018, p.13).

The US tied its food aid in 1965 to a number of economic and political reforms, and even briefly suspended the food aid program over India's conflict with Pakistan (Cleaver 1972; Lerner 2018). This 'dependence upon a Western power's goodwill proved a source of shame for Indian leaders, many of whom fought for decades to wrest their nation from the British empire' (Lerner 2018, p.6). Following Nehru's death in 1964, the new Prime Minister Shastri was also fearful of the effects of US pressure on India's sovereignty. Almost immediately he appointed Subramaniam as Minister for Agriculture and Food, to implement a series of agricultural reforms to cement the 'link between sovereignty and per capita food availability…to achieve food security without Western handouts' (p.10).

Norman Borlaug, who had been a regular visitor to the sub-continent since 1962, flew in 200 kg of Mexican HYV wheat seed each for Pakistan and India, and despite some technical issues the seeds showed great promise and were compatible to local conditions (Paarlberg 1970, p.15). In 1965 India increased its scarce foreign exchange allocation for the purchase of agriculture imports six-fold. Among other things this funded, with the Rockefeller Foundation, a much larger HYV seed import program.

> Additionally, for the 1965–1966 growing season, Subramaniam's ministry launched the National Demonstration Programme, which imported 200 tons of high yield variety wheat seeds from Mexico and distributed them, along with fertiliser, to farmers owning 1,000 well-irrigated plots – the controversial beginning of a targeted agricultural investment policy known as "betting on the strong".
>
> (Lerner 2018, p.10)

Less than a year later, when Indira Gandhi became Prime Minister in 1966, she saw the Green Revolution as 'a question of sheer survival' (quoted in Lerner 2018, p.14). She was under political pressure from her own party on the one hand to respond to the severe drought and an increasing number of food riots, and on the other hand from the US President Johnson who objected to a range of India's policies namely: its population policy; overvalued currency; and its continued criticism of the US war in Vietnam. In response India, again with the help of the Rockefeller Foundation, increased its HYV seed wheat imports from Mexico to 18,000 tonnes, much of which came from Mexican farmers, and that with local plant breeding from the nascent agricultural university system, enabled 700,000 acres to be planted (Paarlberg 1970, p.16). The increase in HYV plantings, together with good rains in 1967, resulted in a dramatic turnaround of India's agricultural fortunes: between 1966 and 1968 grain production increased from 63 million tonnes to 83 million tonnes, and by 1971 it had reached 95 million tonnes (Lerner 2018, p.8). Similar to the Mexico case, wheat was the stand out success, increasing from 12 million tonnes in 1965 to 20 million tonnes in 1970. The success in rice production was more modest and probably due more to weather than HYVs, and it went from 37 to 40 million tonnes (Frankel 1971/2015, p.7).

In some areas yields doubled and income for the peasant farmer increased by 70 per cent. Frankel argues that in the absence of any other supporting policies, HYV wheat needed at least 15 acres of cultivation to be viable under ideal conditions (p.29), thus cutting out the small-scale, marginal, and women farmers completely. This level of landholding to gain the economies of scale from HYVs is probably exaggerated as increased production is generally scale neutral (Freebairn 1995). The real issue is the different risk profiles of larger and smaller farms and women farmers, and their capacity (particularly for women) to access credit for their inputs (Pearse 1980/2015).

The Mexican seed program was aimed at larger farms, which were seen to be better equipped to make full use of the HYVs (Perkins 1990; Lerner 2018, p.10), and at the time this made sense to alleviate a crisis. The targeting of larger farmers unfortunately continues into the 2010s, leading to a 'scale bias in research and extension' (Pingali 2012, p.12304). HYV adoption still has not trickled down to the women, poor and marginal farmers, due to the ongoing lack of government support. Edwin Wellhausen (1976) one of the founders of the Green Revolution, in relation to Mexico's rapid growth in wheat production in the 1950s, also made a similar observation that these programs failed to reach the poor farmers. The positive lessons from Japan in the 1930s, and Europe earlier in reaching out to smallholder farmers were well known in the 1950s and 1960s but ignored in how the Green Revolution was implemented at the time (Hardin 2008; Harwood 2012, 2013; Zanden 1991).

This immediate success and staving off famine gave Indira Gandhi's government the space for a more independent foreign policy *vis à vis* the US and enabled more left leaning popular domestic policies (Das 1999). One of these was to nationalise the private banks to allow, among other things, increased credit being allocated to agriculture. These peasant landowners were also supported with: generous electricity subsidies for pumping water from deep tube wells; expanded canal irrigation; as well as other inputs subsidies to encourage increased productivity (Das 1999). These and other supportive policies facilitated the shift to a landholder capitalist system, based on expanded peasant production, from the existing feudal system. Even as early as 1969 the distributional effects of the Green Revolution were clearly evident as tensions emerged between these land owing peasants and the landless labourers who were not seeing the benefits of these changes with increased wages (Frankel 1971/2015, p.9).

This success of the Green Revolution also changed the political landscape in India, giving medium to large peasant landholders a stronger political voice that ensured the continued protection of the larger scale peasant based agricultural sector, with subsidies arguably at the expense of the poor and marginal farmers (Das 1999; Cleaver 1972; Kaviraj 1988). There was, however, a setback in 1975 when the first oil crisis led to an increase in the price of fertilizer and higher food prices. Headey and Fan (2008) argued that this was one reason for Indira Gandhi's emergency rule in India over the following two years (Headey and Fan 2008). The effect was that the Green Revolution had, and continues to be, hostage to the price of oil (Holt-Giménez and Peabody 2008). This was followed by the winding back of state led research and extension services as a result of neo-liberal economic reforms in the 1990s, which also led to a slowing down in the growth of agricultural productivity (Chadha 2003; Visvanathan 2003). As Visvanathan notes:

> The debates of marginality, equity, vulnerability, the commons, finds no space in this discourse as biotechnology and the liberal imagination combine to create the general property regime. The basic linkages in this world of global linkages ... is between the new economics and genetic science.
> (Visvanathan 2003, p.14)

The role of the state in facilitating the Green Revolution related processes and supporting infrastructure continues to be central to agricultural productivity across India, in much the same way as has occurred in China.

China

China's Green Revolution started at much the same time as its Western counterpart, following immediately the establishment of the Peoples Republic of China (PRC) in 1949; however, it was also quite different to the one led by the US and the Rockefeller Foundation in the rest of Asia in a number of key dimensions. It was more clearly built around complementary policies and state investment that had more marked effects than more liberal based Green Revolution policies elsewhere (Huang and Rozelle 1995; Patel 2013; Rosset, Collins, and Lappé 2000). The state invested heavily in agricultural research, more so than in any other developing country at the time (Fan and Pardey 1997); it managed the development of agriculture infrastructure such as the expansion of irrigation; and it provided inputs such as high yielding seed varieties, including hybrids, as well as fertilizer and pesticides (Moseley 2013). But it was land reform in China that set the scene for the continued success until the 2000s (Fan and Pardey 1997). This is despite the belief that in some places land reform was not necessary for the Green Revolution to be successful (Lerner 2018).

In 1950 the feudal landlord system was reformed over a three-year period, to give small plots of land to the peasants who had to till the land themselves (much like the US and Australia in the nineteenth century), but in China these plots were not transferable. By 1953, 310 million peasants were allocated 47 million hectares of agricultural land plus farming implements and livestock (Lin 2015; Wong 1973). In 1955–1956 this policy changed so that the land was consolidated into collectives (Nolan 1976; Putterman 1988) and later modified in a series of steps from a collective to a commune system in the late 1950s. Finally, in order to maximise output, the policy reverted back to family managed plots in the late 1970s.

The larger communes were a failure due to the centralised policy governing how they functioned. They had unrealistic output targets, large effective taxation by the centre to fund industrialisation and repay Soviet debts, and a poor distribution network for the food the communes produced. The commune system was a part of the Great Leap Forward of rapid industrialisation to break China's dependence on the Soviet Union; however, the result was a famine from 1959 to 1961 in which 20–40 million peasants perished (Yang 1996). This stemmed from the catastrophic failure of the commune system and the policies that guided it (Huang and Rozelle 1995; Manning and Wemheuer 2011).

This calamity led to a refocusing of agricultural policy in the early 1960s to 'ensuring food self-sufficiency became a major priority for the government from that point forward' (Moseley 2013, p.8). The commune system was reformed with the promotion of an 'agriculture first' policy, '[to] intensify production by increasing the use of chemical fertilizer, high-yielding varieties, and water control', which proved successful and provided a spurt to agriculture growth (Huang and Rozelle 1995, p.854). The commune system, however, was progressively wound back from 1978, first with a reversion to the policies of the early 1950s and the introduction of a household production responsibility system. And then, in 1983, the whole commune systems was dismantled in favour of family farms (Lin and Ho 2005; Putterman 1988).

In the 1950s China also rapidly expanded the irrigation system, which probably had its antecedents in work by the Rockefeller Foundation and the Tennessee Valley Authority in the 1930s.

> Throughout the late 1930s and 1940s, the Foundation sent a number of Chinese engineers and agriculturists to Tennessee to witness the accomplishments of the New Deal effort. Rockefeller's example of education (or re-education) of Asians through the model of the TVA was to be repeated with regularity throughout the coming decades.
> (Ekbladh 2002, p.339)

In the 1950s and 1960s the growth in cereal grains in China was mainly due to increased irrigation and land tenure changes. In the 1950s, however, China started a nascent program to breed high yielding hybrid varieties of rice, the main cereal crop. This program came to fruition in the 1970s when HYVs were widely introduced across the country, and alone accounted for 20 per cent of the growth in productivity (Fan and Pardey 1997; Huang and Rozelle 1995; Peng, Tang, and Zou 2009). In addition, the redistribution of commune capital items, such as machinery built up during the Maoist era to family farms also a played part in the increased agricultural output (Moseley 2013).[1] This, combined with households being more responsible for their own production from the late 1970s, not only led to rapid increases in production, but also later the diversification to higher value horticultural and animal products (Fan and Pardey 1997). From 1949 to at least the mid-1990s food production in China on average grew at least one per cent per annum faster than population growth, with the exception of the period of the Great Leap Forward (Fan and Pardey 1997, p.117).

The growth in China's agricultural production since the 1990s, however, has stagnated. This is arguably due to a lack of attention to a faltering institutional framework and support necessary to those left behind after men's rural urban migration (Peng, Tang, and Zou 2009; Schneider 2015; Wu and Ye 2016). The focus on rapid industrialisation and the resultant high levels of men's migration to cities has left mainly women behind to carry the load of the household farm production. This has led to a number of trade-offs that women have to make to balance their domestic workload with their farm responsibilities in the absence of state support. This has included in some areas a reduction of cropping intensity to one crop per year (Wu and Ye 2016), reduced fertilizer application, or decreasing the density of planting (Peng, Tang, and Zou 2009). The more industrial approach of China to its agricultural production is ignoring the small-scale peasant sector, now largely led by women, arguably at some cost. The agricultural policy of 2017 was clearly aimed at larger agricultural producers with a reference to focusing on family farms at a 'moderate scale' (China State Council 2017, p.5).

Investment and support to the smaller scale peasant system could increase the level of productively and overcome the issues that have arisen with poorly supported women-headed households in rural areas. This would require a specific focus on women. While China's gender policy notes the high proportion of women in agriculture, China's agriculture policy refers to support for women in terms of welfare required for a vulnerable group left behind, rather than as part of the production system (China State Council 2015, 2017). This seems to be in response to the adoption by the Chinese government in the 2010s of more Confucian approaches to the national social structure with women being at best a reserve labour force (Joshua 2017), and having the primary, more nurturing role, in the family rather than a productive role as such, which belongs mainly with the man (Fincher 2016a, 2016b; Wilson 2015; Wu and Ye 2016). This is a long way from Mao's oft critiqued dictum that 'women hold up half the sky' (Howell 2002; Maurer-Fazio, Rawski, and Zhang 1999).

Up until the 2000s at least, the more integrated approach to Green Revolution technologies by China, such as integrating infrastructure with the introduction of HYV and hybrid varieties, was arguably more successful than the Western approach of breeding and distributing high yielding seed varieties from centralised research centres. This latter approach was heavily reliant on strong supportive local institutional policies, or the private market in the relevant countries, which were not always available.

Conclusion

These three case studies highlight the tension between 'betting on the strong' at the expense of the small-scale peasant farmers. In the case of India and Mexico it was the larger peasant farms which had access to irrigation, and the focus was mainly on wheat initially, and only later other crops were included that were grown by poorer farmers. In China the focus initially was on feudal land distribution to peasant farmers which was then collectivised into the failed commune farms that were later broken up to peasant farms again. More recently, China policy has been to focus on large farms at the expense of the small-scale farms which are often managed by left behind women after the men have migrated for work. In all of these accounts the role of women as farm managers or principle cultivators of particular crops has been systemically ignored in agricultural research and government policy and is a theme that will be returned to in Chapter 5.

Note

1 For example, cereal production grew at an annual rate of 2.5% from 1952–1978 and 4.7% from 1978–1984, and grain yields grew at 2.8% and 5.9%, respectively over the same periods (Huang and Rozelle 1995, p.853).

References

Abraham, Itty. 2008. "From Bandung to NAM: Non-alignment and Indian Foreign Policy, 1947–1965." *Commonwealth & Comparative Politics* 46(2): 195–219.

Becerril, Javier, and Awudu Abdulai. 2010. "The impact of improved maize varieties on poverty in Mexico: A propensity score-matching approach." *World Development* 38(7):1024–1035.

Berger, Mark T. 2004. "After the Third World? History, destiny and the fate of Third Worldism." *Third World Quarterly* 25(1): 9–39.

Borlaug, Norman E. 2002. *The Green Revolution Revisited and the Road Ahead*. Stockholm: Nobel Prize Stockholm, Sweden.

Brush, Stephen B., Mauricio Bellon Corrales, and Ella Schmidt. 1988. "Agricultural development and maize diversity in Mexico." *Human Ecology* 16(3): 307–328.

Chadha, GK. 2003. "Indian agriculture in the new millennium: Human response to technology challenges." *Indian Journal of Agricultural Economics* 58(1): 1.

China State Council. 2015. *"White Paper": Gender Equality and Women's Development in China*, edited by The State Council Information Office of the People's Republic of China. Beijing: State Council.

China State Council. 2017. *China's Annual Agricultural Policy Goals*, The 2017 No. 1 Document of the CCCPC and the State Council (unofficial translation), edited by Global Agriculture Information Network. Beijing: USDA Foreign Agriculture Service.

Clawson, David L., and Don R. Hoy. 1979. "Nealtican, Mexico: A peasant community that rejected the 'Green Revolution'." *American Journal of Economics and Sociology* 38(4): 371–387.

Cleaver, Harry M. 1972. "The Contradictions of the Green Revolution." *The American Economic Review* 62(1/2): 177–186.

Cullather, Nick. 2010. *The Hungry World: America's Cold War Battle Against Poverty in Asia.* Cambridge (Mass): Harvard University Press.

Das, Raju J. 1999. "Geographical unevenness of India's Green Revolution." *Journal of Contemporary Asia* 29(2): 167–186.

DeWalt, Billie. 1985. "Farming Systems Research: Anthropology, Sociology, and Farming Systems Research." *Human Organization* 44(2): 106–114.

Eddens, Aaron. 2017. "White science and indigenous maize: The racial logics of the Green Revolution." *The Journal of Peasant Studies*: 1–20.

Ekbladh, David. 2002. "'Mr. TVA': Grass-roots development, David Lilienthal, and the rise and fall of the Tennessee Valley Authority as a symbol for US overseas development, 1933–1973." *Diplomatic History* 26(3): 335–374.

Evenson, Robert Eugene, Carl Pray, and Mark W. Rosegrant. 1998. *Agricultural Research and Productivity Growth in India.* Vol. 109: International Food Policy Research Institute.

Fan, Shenggan, and Philip G. Pardey. 1997. "Research, productivity, and output growth in Chinese agriculture." *Journal of Development Economics* 53(1): 115–137.

Fincher, Leta Hong. 2016a. "China's Feminist Five." *Dissent* 63(4): 84–90.

Fincher, Leta Hong. 2016b. *Leftover Women: The Resurgence of Gender Inequality in China.* London: Zed Books.

Frankel, Francine R. 1971/2015. *India's Green Revolution: Economic Gains and Political Costs.* Princeton NJ: Princeton University Press.

Freebairn, Donald K. 1995. "Did the Green Revolution concentrate incomes? A quantitative study of research reports." *World Development* 23(2): 265–279.

Hardin, Lowell S. 2008. "Meetings that changed the world: Bellagio 1969: the green revolution." *Nature* 455(7212): 470.

Harwood, Jonathan. 2012. *Europe's Green Revolution and Others Since: The Rise and Fall of Peasant-Friendly Plant Breeding.* Abingdon: Routledge.

Harwood, Jonathan. 2013. "Has the Green Revolution been a cumulative learning process?" *Third World Quarterly* 34(3): 397–404.

Headey, Derek, and Shenggen Fan. 2008. "Anatomy of a crisis: The causes and consequences of surging food prices." *Agricultural Economics* 39(s1): 375–391.

Hill, Brent. 2017. "The Green Revolution and the Rockefellers." *Isocraccy*, Feb. 28.

Holt-Giménez, Eric, and Loren Peabody. 2008. "From food rebellions to food sovereignty: Urgent call to fix a broken food system." *Food First Backgrounder* 14(1).
Howell, Jude. 2002. "Women's political participation in China: Struggling to hold up half the sky." *Parliamentary Affairs* 55(1): 43–56.
Huang, Jikun, and Scott Rozelle. 1995. "Environmental stress and grain yields in China." *American Journal of Agricultural Economics* 77(4): 853–864.
Joshua, John. 2017. "Balanced Path Development" in *China's Economic Growth: Towards Sustainable Economic Development and Social Justice*, Vol 2. London: Palgrave McMillan, pp.13–45.
Kaviraj, Sudipta. 1988. "A critique of the passive revolution." *Economic and Political Weekly*: 2429–2444.
Kilby, Patrick. 2017. *China and the United States as Aid Donors: Past and Future Trajectories, Policy Studies 77*. Honolulu: East West Center.
Lerner, Adam B. 2018. "Political Neo-Malthusianism and the progression of India's Green Revolution." *Journal of Contemporary Asia* 48(3): 1–23.
Lin, Chun. 2015. "Rethinking land reform: comparative lessons from China and India." In *The Land Question: Socialism, Capitalism and the Market.*, edited by Mahmood Mamdani, pp. 95–157. Kampala: Makerere Institute of Social Research
Lin, George, and Samuel P.S. Ho. 2005. "The state, land system, and land development processes in contemporary China." *Annals of the Association of American Geographers* 95(2): 411–436.
Loveridge, Jack. 2017. "Between hunger and growth: Pursuing rural development in Partition's aftermath, 1947–1957." *Contemporary South Asia* 25(1): 56–69.
Manning, Kimberley Ens, and Felix Wemheuer. 2011. *Eating Bitterness: New Perspectives on China's Great Leap Forward and Famine*. Vancouver: UBC Press.
Maurer-Fazio, Margaret, Thomas G. Rawski, and Wei Zhang. 1999. "Inequality in the rewards for holding up half the sky: Gender wage gaps in China's urban labour market, 1988–1994." *The China Journal* (41): 55–88.
Moseley, William G. 2013. "The evolving global agri-food system and African–Eurasian food flows." *Eurasian Geography and Economics* 54(1): 5–21.
Niazi, Tarique. 2004. "Rural poverty and the Green Revolution: The lessons from Pakistan." *The Journal of Peasant Studies* 31(2): 242–260.
Nolan, Peter. 1976. "Collectivization in China: Some comparisons with the USSR." *The Journal of Peasant Studies* 3(2): 192–220.
Paarlberg, Don. 1970. *Norman Borlaug, Hunger Fighter*. Washington: Foreign Economic Development Service, US Department of Agriculture.
Paddock, William, and Paul Paddock. 1967. *Famine-1975! America's Decision: Who Will Survive?* Boston: Little, Brown and Co.
Patel, Raj. 2013. "The long green revolution." *The Journal of Peasant Studies* 40(1): 1–63.

Pearse, Andrew. 1980/2015. "Seeds of plenty, seeds of want: Social and economic implications of the Green Revolution." In *Revisiting Sustainable Development 2015* (Reprinted), edited by Peter Utting. Geneva: UNRISD

Peng, Shaobing, Qiyuan Tang, and Yingbin Zou. 2009. "Current status and challenges of rice production in China." *Plant Production Science* 12(1): 3–8.

Perkins, John H. 1990. "The Rockefeller Foundation and the green revolution, 1941–1956." *Agriculture and Human Values* 7(3–4): 6–18.

Phillips, Ronald L. 2013. 'Norman Ernest Borlaug. 25 March 1914–12 September 2009.' In *Biographical Memoirs of Fellows of the Royal Society*. London: The Royal Society.

Pingali, Prabhu L. 2012. "Green Revolution: Impacts, limits, and the path ahead." *Proceedings of the National Academy of Sciences* 109(31): 12302–12308.

Putterman, Louis. 1988. "Group farming and work incentives in collective-era China." *Modern China* 14(4): 419–450.

Rosset, Peter, Joseph Collins, and Frances Moore Lappé. 2000. "Lessons from the Green Revolution." *Third World Resurgence*: 11–14.

Schneider, Mindi. 2015. "What, then, is a Chinese peasant? Nongmin discourses and agroindustrialization in contemporary China." *Agriculture and Human Values* 32(2): 331–346.

Smale, Melinda, Mauricio R. Bellon, and Jose Alfonso Aguirre Gomez. 2001. "Maize diversity, variety attributes, and farmers' choices in Southeastern Guanajuato, Mexico." *Economic Development and Cultural Change* 50(1): 201–225.

Sonnenfeld, David A. 1992. "Mexico's 'Green Revolution', 1940–1980: Towards an environmental history." *Environmental History Review* 16(4): 29–52.

Sumberg, James, Dennis Keeney, and Benedict Dempsey. 2012. "Public agronomy: Norman Borlaug as 'brand hero' for the Green Revolution." *The Journal of Development Studies* 48(11): 1587–1600.

Tan, Tai Yong, and Gyanesh Kudaisya. 2008. *Partition and Post-Colonial South Asia: A Reader*. London: Routledge.

Tuckman, Barbara H. 1976. "The green revolution and the distribution of agricultural income in Mexico." *World Development* 4(1): 17–24.

USAID. 1982. "Administrator's International Development Leaders Forum: Norman Borlaug." In *Norman E. Borlaug Background Material, March 26*, edited by USAID. Washington: USAID.

Visvanathan, Shiv. 2003. "From the Green Revolution to the Evergreen Revolution: Studies in Discourse Analysis." IDS Seminar on Agriculture Biotechnology and the Developing World. October, Brighton, IDS.

Wellhausen, Edwin J. 1976. "The agriculture of Mexico." *Scientific American* 235(3): 128–153.

Wilson, Japhy. 2015. "A strange kind of science: Making sense of the millennium villages project." *Globalizations* 12(4): 645–659.

Wong, John. 1973. *Land Reform in the People's Republic of China. Institutional Transformation in Agriculture*. New York: Praeger
Wu, Huifang, and Jingzhong Ye. 2016. "Hollow lives: Women left behind in rural China." *Journal of Agrarian Change* 16(1): 50–69.
Yang, D.L. 1996. *Calamity and Reform in China: State, Rural Society, and Institutional Change since the Great Leap Famine*. New York: Stanford University Press.
Zanden, Jan. 1991. "The first green revolution: The growth of production and productivity in European agriculture, 1870–1914." *The Economic History Review* 44(2): 215–239.

4 Countervailing forces
Structural adjustment and the twenty-first century Green Revolution

Introduction

The three case studies from Chapter 3 make the point that the success of the Green Revolution was dependent on a clear policy framework in which land reform with a focus on the peasant farmer was a central element. Many of the complementary policies to support the uptake of Green Revolution technologies were not only absent, but actively discouraged by often contradictory and ideologically driven foreign aid policies of the West (Blackden and Morris-Hughes 1993; Fonchingong 1999; Koehler 2015). A key policy of structural adjustment was to reduce the size of government, which meant cutting state expenditure on supporting rural infrastructure such as roads, marketing services, and agricultural extension, which as we have seen, is necessary for the successful uptake of many if not most Green Revolution technologies. In the 1980s there was a move to a farming systems approach to agricultural research which was redolent of the late nineteenth and early twentieth century approaches (Harwood 2012). But this was at odds with prevailing neo-liberal policy and approaches promoted by the World Bank and many Western donors. These favoured reduced government investment in agricultural support services, cash crop production for export, and larger commercial farms. The search for the 'quick fix' rather than the long term payoff of a farming systems approach resulted in the waning of donor interest in the Green Revolution.

Donors led by the World Bank were opposed to increasing agricultural productivity through state investment in agriculture, despite its success in a number of countries, not least of which China. They believed that investment in agriculture should be left to market forces, eschewing policies that promoted cross economy transfers to support particular sectors. Africa, more aid dependent, was subject to structural adjustment policies based on a philosophy that agricultural policies should be market based, even in places where there were no viable markets. Thus, the 1980s became known as the lost decade in Africa's agricultural development, where investment lagged, and rural areas were neglected (Carrasco 1999; Edwards 2013; Edwards 1989). In Asia, Cold War contestations among donors and geo-political imperatives, such as the opening up of China, reduced the impacts of these adverse, more ideologically based, donor policies.

China was effectively exempted from these structural adjustment policy conditions for World Bank loans, as China was a high priority for World Bank lending to the point that the Bank generally acquiesced to China's terms (World Bank 1981, 1984, 1985). This, in addition to large amounts of assistance from Japan (Smith 2015), enabled China to continue the rapid pace of agricultural and industrial development into the 2000s with little interference from the outside.

The Green Revolution in the 2000s

The so-called Second Green Revolution of the early 2000s, which in this narrative is more correctly a third Green Revolution, mirrors some of the Cold War rivalries of the 1950s. The rising influence of China in the agriculture space, particularly in Africa (Bräutigam and Xiaoyang 2009), and the importance of the West to remain relevant in the developing world may be the basis of this rivalry. The key difference is that the rivalry is not explicitly stated. While there is some concern about China's role in aid and development (Bräutigam 2011; Bräutigam and Xiaoyang 2009; Bräutigam and Xiaoyang 2016), the idea of the US matching it is generally unstated.

The twenty-first century Green Revolution has a strong, but not exclusive focus on Africa. The focus on Africa is in part to fill the gaps from the post-war Green Revolution, and, in part, of seeing Africa as a source of land for more intensive large-scale cultivation, and the associated corporatisation of African agriculture. The Green Revolution in Africa takes a number of forms: from improving fodder for livestock; multi-cropping; the continued development of HYVs for common agricultural crops such as sorghum and millet; and the introduction of HYVs from abroad such as Chinese hybrid rice varieties, and GMOs (genetically modified organisms) from the West. Much like HYV varieties of wheat that were introduced into India in the 1960s, now the HYVs for Africa are in the form of rice from China. The difference is that rice is much less an important crop in Africa than wheat was in India, so there are many more barriers to its adoption.

Many countries have also been involved in what are referred to as 'land grabs' whereby foreign interests are buying increasingly larger swathes of land for agriculture across Africa (Bräutigam and Xiaoyang 2009; Moyo 2016). This has led to

> the marginalisation of family farming and the proletarisation of farmers, who are becoming rent-seekers or landless labourers. Even where the land rush could lead to increased productivity, there are real concerns that the most vulnerable groups will bear the costs without reaping the benefits and that the host economies are not effectively benefiting from it.
>
> (Anseeuw 2013, p.172)

There is a constellation of interests among donors in agricultural development in Africa. These range from supporting large scale corporate farming to supporting small-scale and marginal farmers, and both national and private corporate interests seeking to introduce their new HYVs to licence and sell to local farmers (Azadi et al. 2015; Azadi et al. 2017).

China and Africa

In the late 1990s the Millennium Partnership for African Recovery (MPA) was put in place, initially for debt relief, but became a more encompassing New Economic Partnership for African Development (NEPAD) in 2001, which included the African Union and Western donors as well as China 2001 (Akinola and Ndawonde 2016; Hamad and Kitigwa 2016). The Forum on China-Africa Cooperation (FOCAC), in a sense seemed to have anticipated NEPAD, having been established a year earlier in October 2000. FOCAC is a series of triennial ministerial level conferences to oversee and report on the progress of Chinese aid and investment across Africa, some of which would be in agriculture (Shelton and Paruk 2008; Taylor 2010; Tugendhat and Alemu 2016). The 2015 FOCAC meeting had a specific focus on Agriculture and Food Security (FOCAC 2015), with the main activities being technical exchanges and agricultural demonstration centres. The 2018 meeting reiterated the focus on rural socio-economic development, and gender equality (FOCAC 2018). Whether gender equality is translated into agricultural research and development programs remains to be seen.

Much of the aid and investment from China has been in infrastructure to lay the foundation for further agricultural and industrial development. This is in line with the New Structural Economics theories, which emphasise the role of infrastructure in achieving sustainable growth (Carey and Xiaoyun 2016; Lin and Wang 2014) and has always had a program of direct investment in agriculture as well. As noted above in the 2015–2018 plan it has been enhanced and made a priority (Besada and O'Bright 2017; Moyo 2016; Mthembu 2016; Sun 2015).

China, like other donors, has struggled over the past sixty years to make a lasting and sustainable impact in its foreign aid of agricultural development and food security (Bräutigam and Xiaoyang 2009). From the state-owned farms of the 1960s to the demonstration centres from the late 1970s, all have struggled to be sustainable with the quality of the Chinese experts and their capacity to work with local communities being seen as part of the problem (Bräutigam and Xiaoyang 2009; Xu et al. 2016). The solution however is not clear: one response has been to increase Chinese management and control of these demonstration centres to the point they are seen as being Chinese owned and run farms. Many of the large farms run by Chinese to export produce back to China have experienced a local backlash to the point that the China run and managed approach of some of these large Chinese owned farms is limited, as China regards its positive reputation as paramount.

> Technologies therefore travel from China to Africa not just as "things", but they are bound up with social, historical, and political meanings and implications. This reflects the ideological and political dimensions of technology transfer agricultural development cooperation. Technologies, initially constructed in particular Chinese settings, travel with these contexts, and so adopt a particular "rationality", that is at once technical, social, cultural and political, and embedded in historical experiences.
>
> (Xu et al. 2016, p.84)

The Chinese agricultural program, like those of the West, has also been accused of preferring use of Chinese HYV seeds to expand China's market, rather than meeting farmer needs (Bräutigam and Xiaoyang 2009; Xu et al. 2016). For example, across Africa, China has set up 'demonstration centres' that are meant to be self-funding after around five years. These have a historical antecedence dating back to Sun Yat Sen in 1906 and an ideology of 'technocratic rationality', when he set up demonstration farms across China, with a focus on 'food security through modern agricultural technology' (Xu et al. 2016, p.83). This was through a top-down technocratic approach, but in twenty-first-century Africa the focus is on commercial advantage as well, as this example from a Tanzanian agricultural development centre demonstrates:

the Chinese team leader [Mr. C] was more optimistic: "Rice varieties are the core competitive products of our research companies. Since we are here to do the demonstration, we need to demonstrate the most advantageous varieties we have at hand. We actually do not have much technical advantage on maize production." The logic behind the technology selection is clear: hybrid rice is the product that they hope to present and to sell, and the business venture of the ATDC [Agricultural Technology Demonstration Centers] is highly dependent on a market demand being created and demonstration? Mr. C explained: "Rice varieties are the core competitive products of our research companies."

(p.87)

While there were some successes in the 1980s and 1990s with mixed private public models, the more recent models are more commercial in focus. This often involves exporting Chinese agricultural technology, but now based around rice, to communities where maize is the main crop. Like many western HYVs, not only are these HYV rice varieties not relevant, but without the supporting agronomic requirements, the poor farmer cannot afford them (Xu et al. 2016, p.86). Likewise, if these 'demonstration farms' had more Chinese management and control then it might lead to a situation where 'large scale production could push people off their own land, using them as seasonal workers' (Bräutigam and Xiaoyang 2009, p.706). While China is very sensitive to these issues at FOCAC level 'there is little evidence that Chinese experts are even aware of these cultural production issues or that they can take steps to ameliorate the impact' (ibid.). This is all too common a story from the post-war Green Revolution in Asia and Latin America, and it seems to be being repeated in Africa and not only by China.

The West and Africa

FOCAC of course was not the only game in town, and may have been part of an unstated 'rivalry' between China and the West (Kilby 2017; Moseley, Schnurr, and Bezner Kerr 2015). There was and continues to be a veritable network of interlocked programs as various Western and other interests tried to get a foot in the door to what was seen as a new frontier for technical development, or more cynically neo-colonial exploitation of Africa. FOCAC was to become part of NEPAD, but also supporting NEPAD was a UN led initiative, the Comprehensive Africa Agricultural Development Program (Kolavalli et al. 2010), and the Africa Green Revolution established in 2004 by UN Secretary general Kofi Annan (Annan 2004; Diao, Headcy, and Johnson 2008). This latter program has been heavily supported by the West under the Alliance for Green Revolution in Africa (AGRA). Like the first Green Revolution the Rockefeller Foundation, together with the Bill and Melinda Gates Foundation, have been deeply involved in AGRA (Ejeta 2010, p.832). These two Foundations have provided support for a secretariat, and over $300 million in start-up funds for seed and soil health programs (Toenniessen, Adesina, and DeVries 2008, p.240). They aim to integrate

> the most prosperous smallholders into the singular global market; and coordination of food policies within regions of Africa…to link African food production and consumption into the global food chain, controlled by a cartel of very few corporations.
> (Thompson 2012, p.345)

AGRA finances research and production of commercial HYV seeds (including GMOs) under licence; this expands their market, and ties the local farmers to a commercial arrangement, along with the necessary components of fertilizers and pesticides. AGRA has a clear similarity to the post-war Green Revolution in that it focuses on larger peasant farmers and ties them into capital expenditure not only on fertilizers and pesticides but also genetically modified or hybrid seeds, neither of which can be reproduced at farm level. While the success of these processes is nowhere near as dramatic as some of the earlier cases from India and Mexico, the increase in inequality the earlier Green Revolution brought is being replicated with these programs in Africa (Patel 2013; Shilomboleni 2017).

Then there is the New Alliance for Food Security and Nutrition (NAFSN), a US led Group of Eight initiative launched in 2012 (Moseley, Schnurr, and Bezner Kerr 2015; Patel et al. 2015;). This is private industry and Western technology based, and like AGRA is much less about the poor farmer and basic nutrition issues, despite its title, than corporate entry into a new market (Moseley, Schnurr, and Bezner Kerr 2015; Patel et al. 2015; Vercillo et al. 2015).

> Given the close links between this initiative and the business community, one could argue that those in the American aid circles have learned from the Chinese whose development assistance has very tight links to Chinese business interests.
> (Moseley, Schnurr, and Bezner Kerr, p.5)

NAFSN has at its core a regulatory environment through Country Cooperation Frameworks that protects commercial seed providers from local competition through intellectual property and regulation processes, and locks farmers into long term production contracts and the associated risks that come with them (De Schutter 2015).

In contrast to the commercial focus of China and the US in promoting HYVs through FOCAC, AGRA, and NAFSN, the US has also invested heavily in the Feed the Future program with 24 Innovation Labs based in US universities. These target poor and marginal farmers in food production and improved nutrition with farmer led technology. This is a follow-on program with the Cooperative Research Support Program (CRSP) that dates back to the 1970s (Weir and Williamson 2017). These Innovation Labs cover the full gamut of agriculture research ranging from sustainable intensification to dealing with post-harvest issues, integrated pest management, small-scale irrigation, and a broad range of crops, as well as livestock, and aquaculture and fisheries, all in the context of climate change (Du et al. 2015; Ho and Hanrahan 2011; Prasad and Middendorf 2016).

This is one of the few larger programs that moves away from the more corporate approaches supported by the Chinese government on the one hand and the Rockefeller Foundation on the other. The Feed the Future program has a specific focus on technology to support existing crops and livestock grown by the poor and marginal farmer and expand them on a systems basis (Olewnik and Hackett 2017; Tittonell 2014). The lesson it learnt from the CRSP program is the importance of local research partnerships; however, it still faces the challenge of incorporating the social sciences research on farming systems approaches and in particular the role of women as farmers or principle cultivators of particular crops. This contrasts with, for example, the Millennium Villages Program that also espouses a village systems approach but is still driven by externally set targets, and a one size fits all approach.

Millennium Villages Program

The Millennium Villages Program (MVP) is another approach to the Green Revolution to try to overcome some of the failures of the past and focus on poor farmers, but ultimately it is caught up in the same issues that bedevil most outsider driven development projects aimed at introducing new technology. The MVP is a high-profile program initiated by Prof. Geoffrey Sachs and the Earth Institute at Colombia University in which 80 model villages were set up in 11 clusters of a total of 80 villages across Sub-Saharan Africa funded at a level of $120 per person per year (Kimanthi and Hebinck 2018; Wilson 2017). It has been described as a

> prime example of social engineering of a "model village-style social experiment" and again as "a living laboratory" whereby massive investments are made in integrated programmes at village level through planned interventions within a specific timeframe.
>
> (Kimanthi and Hebinck 2018, p. 157)

The MVP harks back to some of the failed social experiments of the 1950s and 1960s of alternative lifestyles, and some state programs in the Soviet Union and China. In this case the goals are not about an alternative lifestyle, or meeting national production targets, but in meeting the Sustainable Development Goals (SDGs). The MVP was aimed at applying Green Revolution technologies of improved crops and associated support, but in an integrated development environment of improving other development factors such as health, education, communications, safe water etc. In practice the program provided 'free' packages of fertilizers and seeds for HYVs crops as well as fallow crops to farmers. These farmers were assumed to be homogenous, devoid of wealth, gender, religion, or ethnic diversity. These packages were conditional on farmers organising themselves in certain ways, and that after some time the free input would be phased out.

In practice the program, being heavily top-down, did not run as planned from the outset with some farmers selling the inputs to neighbouring areas rather than carrying the risks of their own crop failure, farmers being unwilling to organise in the ways suggested, and a reluctance to repay the loans prompting coercion in some places.

> The debt collectors, who were also members of the community, would take it upon themselves to assess the farmer's harvest from the farmers' stores and extract the number of bags indebted as the cooperative input loan. They would force their way in whenever a farmer tried to resist.
>
> (Kimanthi and Hebinck 2018, p.165)

This led to a fairly common set of problems relating to these types of top-down interventions: that is, favouritism and elite capture, among other things, all leading to increased inequality and broken communities. Wilson (2015) puts it into the realm of fantasy.

MVP is better understood not as a strategy of social engineering but as the staging of a social fantasy in which the constitutive antagonisms of capitalist development are concealed. In this fantasy space, the MDGs function to fill the gap between the promise of development and the impossibility of capitalism without poverty.

(p.645)

What has characterised all of these approaches discussed above, as part of the twenty-first-century green revolution and how it has played out in Africa, is what seems to be a lack of coordination among the players, more than a certain amount of self-interest, and an active avoidance of the voices of the poor, of women, their interests, and the contexts in which these approaches have been set. Ignoring local context seems to be the common denominator among the range of interventions with the exception of the Feed the Future program that is more cognisant of local context due mainly to its 40 years of cumulative experience. AGRA's approach on the other hand in promoting GMOs with licences from private corporations seems to be in direct competition with similar Chinese programs promoting China's own hybrid varieties (Moseley, Schnurr, and Bezner Kerr 2015; Xu et al. 2016). The sad irony with the AGRA program is that much of the genetic material for the GMOs was originally collected in Africa but the source communities have not been recompensed for that genetic material (Thompson 2012). NAFSN, likewise, is seen as a path for Western agro-capital to invest in Africa, with little to do either with food security or nutrition. Overall, the effect of all of these interventions has been to marginalise poor peasant and women farmers further (Moseley, Schnurr, and Bezner Kerr 2015; Patel et al. 2015; Vercillo et al. 2015).

What is emerging, particularly in Africa in the 2010s, is a repetition of the mistakes of the First Green Revolution. The promotion of technologies from outside with commercial or other self-interested benefits, without looking at what is needed, or the impact on local populations, often leads to patchy results at best, and at worse wasted resources and failure. This new intellectual property rights regime can also limit researchers' and farmers' access to certain genetic material they most need, as we have seen with the Western and Chinese initiatives discussed above, and so affect the direction research takes (Welch et al. 2017).

Nutrition

While there is no doubt that the Green Revolution increased total food output and assured food supply at least at a national level, the issue of whether it has improved household nutrition is more difficult to generalise about and any data can be misleading (Negin et al. 2009; Pinstrup-Andersen and Hazell 1985). The nutrition levels of say women and children are more dependent on intra-household distribution, and whether new varieties lead to crop substitution at the local level. For example, in India as with other places, the move to monocropping and increased wheat production was often at the expense of pulses and beans (Harwood 2012). The effect was that while total calorie intake increased it was at the expense of the availability of a broad range of nutritional foods (Negin et al. 2009; Pinstrup-Andersen and Hazell 1985), micronutrients found in traditional crops, fish in rice paddies, and wild leafy vegetables are now less commonly available leading to poor health outcomes (Negin et al. 2009; Pingali 2012). While a solution might be to breed micronutrients into the new crops (e.g. the GMO Golden Rice), this is not only playing catch up, it also ties the farmer to GMOs and the costs and risks associated with constantly buying seeds and other inputs. It is not a comprehensive solution, and often ignores the broader systems in which nutritional practices are set. Moving away from monocrops and integrating secondary crops and intercropping is a solution that some research looks into, but these voices are often drowned out by corporate interests.

New technology and local communities

One theme that emerges from the Green Revolution in its various iterations is the extent to which new technology being applied connects with local communities and their needs. In the 1870s to the 1920s European Green revolution, and its counterparts in places like Australia, the US, and Japan, the success was mainly in those countries that had a suite of state led support mechanisms including agricultural extension, rural credit, and the supporting physical infrastructure, such as improved roads and rail (Harwood 2012). Those areas that did not have these enabling institutions could not make full use of a particular technology. Similarly, with the post-war Green Revolution those countries or regions that had the strongest supporting institutions, such as China and parts of Mexico and India, were most successful.

The same set of socio-economic conditions in rural areas that occurred in Europe in the late nineteenth century is emerging across Asia and Africa. There is a rural labour shortage as a result of seasonal migration, which has disrupted the traditional family farm to the point that it is now more often the women who is the principle cultivator, at the very least for some time of the year. However, including women in rural development and the changes to farming technologies has been an ongoing sticking point in agricultural innovation for the past 70 years if not longer. The move to the family farm as the locus of production has been a logical outcome of meeting rural labour shortages in farming. The next step is to recognise the gendered nature of the family farm and peasant agriculture more generally, and the central role women play, but this is something that the institutional frameworks of the Green Revolution have not caught up with, and is the focus of Chapter 5.

References

Akinola, Adeoye O., and Nompumelelo Ndawonde. 2016. "NEPAD: Talking from the South, governing from the West." *International Journal of African Renaissance Studies – Multi-, Inter- and Transdisciplinarity* 11(2): 38–51.

Annan, Kofi. 2004. "Africa's Green Revolution—A call to action." Address to the Highlevel Seminar on Innovative Approaches to Meet the Hunger Millennium Development Goal in Africa, July, Addis Ababa.

Anseeuw, Ward. 2013. "The rush for land in Africa: Resource grabbing or green revolution?" *South African Journal of International Affairs* 20(1): 159–177.

Azadi, Hossein, Mansour Ghanian, Omid M. Ghoochani, Parisa Rafiaani, Clauvis N.T. Taning, Roghaye Y. Hajivand, and Thomas Dogot. 2015. "Genetically modified crops: Towards agricultural growth, agricultural development, or agricultural sustainability?" *Food Reviews International* 31(3): 195–221.

Besada, Hany, and Ben O'Bright. 2017. "Maturing Sino–Africa relations." *Third World Quarterly* 38(3): 655–677.

Blackden, C. Mark, and Elizabeth Morris-Hughes. 1993. "Paradigm postponed: Gender and economic adjustment in Sub-Saharan Africa." In *World Bank Group Archives*, edited by World Bank. Washington: World Bank. Technical department Africa region. Human resources and poverty division Technical Note 13.

Bräutigam, Deborah. 2011. "Aid 'with Chinese characteristics': Chinese foreign aid and development finance meet the OECD–DAC aid regime." *Journal of International Development* 23(5): 752–764.

Bräutigam, Deborah A., and Tang Xiaoyang. 2009. "China's engagement in African agriculture: 'Down to the countryside'." *The China Quarterly* 199: 686–706.

Carey, R., and L. Xiaoyun. 2016. "China's comprehensive strategic and cooperative partnership with Africa." *IDS Policy Briefing*, February (111).

Carrasco, Enrique R. 1999. "The 1980s: The debt crisis and the lost decade of development." *Transnational Law & Contemporary Problems* 9: 119.

De Schutter, Olivier. 2015. *The New Alliance for Food Security and Nutrition in Africa*. Edited by Directorate-General for External Policies Policy Department. Brussels European Parliament Committee on Development.

Diao, Xinshen, Derek Headey, and Michael Johnson. 2008. "Toward a green revolution in Africa: What would it achieve, and what would it require?" *Agricultural Economics* 39 (s1): 539–550.

Du, Lidan, Victor Pinga, Alyssa Klein, and Heather Danton. 2015. "Leveraging agriculture for nutrition impact through the feed the future initiative." *Advances in Food and Nutrition Research* 74: 1–46.

Edwards, Michael. 2013. *Future Positive: International Co-operation in the 21st Century.* London: Earthscan.

Ejeta, Gebisa. 2010. "African Green Revolution needn't be a mirage." *Science* 327(5967): 831–832.

FOCAC. 2015. *The Forum on China-Africa Cooperation Johannesburg Action Plan (2016–2018).* Edited by FOCAC. Johannseburg: FOCAC.

FOCAC. 2018. *The Forum on China-Africa Cooperation Beijing Action Plan (2019–2021).* Edited by FOCAC. Beijing: FOCAC.

Fonchingong, Charles. 1999. "Structural adjustment, women, and agriculture in Cameroon." *Gender & Development* 7(3): 73–79.

Hamad, S.O., and M.M. Kitigwa. 2016. "What is new in the New Partnership for Africa's Development (NEPAD)?" *Huria: Journal of the Open University of Tanzania* 23(1): 114–129.

Harwood, Jonathan. 2012. *Europe's Green Revolution and Others Since: The Rise and Fall of Peasant-Friendly Plant Breeding.* Abingdon: Routledge.

Ho, Melissa D., and Charles E. Hanrahan. 2011. *The Obama Administration's Feed the Future Initiative.* Washington: Congressional Research Service.

Ignatova, Jacqueline A. 2017. "The 'philanthropic' gene: Biocapital and the new green revolution in Africa." *Third World Quarterly* 38(10): 2258–2275.

Kilby, Patrick. 2017. *China and the United States as Aid Donors: Past and Future Trajectories, Policy Studies 77.* Honolulu: East West Center.

Kimanthi, Hellen, and Paul Hebinck. 2018. "'Castle in the sky': The anomaly of the millennium villages project fixing food and markets in Sauri, western Kenya." *Journal of Rural Studies* 57: 157–170.

Koehler, Gabriele. 2015. "Seven decades of 'development', and now what?" *Journal of International Development* 27(6): 733–751.

Kolavalli, Shashidhara, Kathleen Flaherty, Ramatu Al-Hassan, and K. Owusu Baah. 2010. "Do Comprehensive Africa Agriculture Development Program (CAADP) processes make a difference to country commitments to develop agriculture? The case of Ghana." In *The Case of Ghana*, edited by The International Food Policy Research Institute (IFPRI). Washington: The International Food Policy Research Institute (IFPRI)

Lin, Justin Yifu, and Yan Wang. 2014. "China-Africa co-operation in structural transformation: Ideas, opportunities, and finances." WIDER Working Paper. 2014/046 Helsinki: UNU-WIDER.

Moseley, William, Matthew Schnurr, and Rachel Bezner Kerr. 2015. "Interrogating the technocratic (neoliberal) agenda for agricultural development and hunger alleviation in Africa." *African Geographical Review* 34(1): 1–7.

Moyo, Sam. 2016. "Perspectives on South–South relations: China's presence in Africa." *Inter-Asia Cultural Studies* 17(1): 58–67.
Mthembu, Philani. 2016. "Reflecting on the Johannesburg Summit of the Forum on China-Africa Cooperation (FOCAC): Where to from here?" In *Global Insight*. Pretoria, South Africa: Institute for Global Dialogue (IGD) and Friedrich-Ebert-Stiftung (FES).
Negin, Joel, Roseline Remans, Susan Karuti, and Jessica C. Fanzo. 2009. "Integrating a broader notion of food security and gender empowerment into the African Green Revolution." *Food Security* 1(3): 351–360.
Olewnik, M., and J. Hackett. 2017. "Kansas State University's collaborative approach to research through global food systems." *Cereal Foods World* 62(6): 267–271.
Patel, Raj. 2013. "The long green revolution." *The Journal of Peasant Studies* 40(1): 1–63.
Patel, Raj, Rachel Bezner Kerr, Lizzie Shumba, and Laifolo Dakishoni. 2015. "Cook, eat, man, woman: Understanding the New Alliance for Food Security and Nutrition, nutritionism and its alternatives from Malawi." *The Journal of Peasant Studies* 42(1): 21–44.
Pingali, Prabhu L. 2012. "Green Revolution: Impacts, limits, and the path ahead." *Proceedings of the National Academy of Sciences* 109(31): 12302–12308.
Pinstrup-Andersen, Per, and Peter B.R. Hazell. 1985. "The impact of the Green Revolution and prospects for the future." *Food Reviews International* 1(1): 1–25.
Prasad, P.V., and B. Jan Middendorf. 2016. "USAID Feed the Future Innovation Lab: Global research in sustainable intensification." 20 July 2016 – ASABE Annual International Meeting Orlando, Florida: Kansas State University.
Shelton, Garth, and Farhana Paruk. 2008. "The Forum on China-Africa cooperation: A strategic opportunity." *Institute for Security Studies Monographs 2008* (156): 222.
Shilomboleni, Helena. 2017. "A sustainability assessment framework for the African green revolution and food sovereignty models in southern Africa." *Cogent Food & Agriculture* 3(1): 1–17.
Singer, Hans Wolfgang. 1989. "The 1980s: A lost decade – development in reverse?" In *Growth and External Debt Management*, 46–56. New York: Springer.
Smith, S.A. 2015. *Intimate Rivals: Japanese Domestic Politics and a Rising China*. New York: Columbia University Press.
Sun, Yun. 2015. "The sixth forum on China-Africa cooperation: New agenda and new approach?" In *Foresight Africa: Top Priorities for the Continent in 2015*, edited by Brookings Growth Initiative, 10. Washington: Brookings.
Taylor, Ian. 2010. *The Forum on China-Africa Cooperation (FOCAC)*. London: Routledge.

Thompson, Carol B. 2012. "Alliance for a Green Revolution in Africa (AGRA): Advancing the theft of African genetic wealth." *Review of African Political Economy* 39(132): 345–350.

Tittonell, Pablo. 2014. "Ecological intensification of agriculture–sustainable by nature." *Current Opinion in Environmental Sustainability* 8: 53–61.

Toenniessen, Gary, Akinwumi Adesina, and Joseph DeVries. 2008. "Building an alliance for a green revolution in Africa." *Annals of the New York Academy of Sciences* 1136(1): 233–242.

Tugendhat, Henry, and Dawit Alemu. 2016. "Chinese agricultural training courses for African officials: Between power and partnerships." *World Development* 81: 71–81.

Vercillo, Siera, Vincent Z. Kuuire, Frederick Ato Armah, and Isaac Luginaah. 2015. "Does the New Alliance for Food Security and Nutrition impose biotechnology on smallholder farmers in Africa?" *Global Bioethics* 26(1): 1–13.

Weir, Collin C., and HandyWilliamson, Jr. 2017. "International involvement of historically black land-grant institutions in AID-supported development activities and programs." In *A Century of Service*, 69–92. Abingdon: Routledge.

Welch, Eric W., Federica Fusi, Selim Louafi, and Michael Siciliano. 2017. "Genetic resource policies in international collaborative research for food and agriculture: A study of USAID-funded innovation labs." *Global Food Security* 15(December): 33–42.

Wilson, Japhy. 2015. "A strange kind of science: Making sense of the millennium villages project." *Globalizations* 12(4): 645–659.

Wilson, Japhy. 2017. "Paradoxical utopia: The millennium villages project in theory and practice." *Journal of Agrarian Change* 17(1): 122–143.

World Bank. 1981. "Memo: from Koch-Weser for record Oct 14. China Mr Clausen meeting with Chinese delegation." In *Country Files China A* A1990–013 1774658, edited by The World Bank. Washington: World Bank Group Archives.

World Bank. 1984. "Koch Weser to file Jan 11: Mr Clausen's meeting with Premier Zhao Ziyang." In *Country Files China A* A1990–013 1774658, edited by The World Bank. Washington: World Bank Group Archives.

World Bank. 1985. "Memo: from Katz on the discussion draft of China paper 25 Sept." In A1994–022#42 edited by World Bank. Washington: World Bank Group Archives.

Xu, Xiuli, Xiaoyun Li, Gubo Qi, Lixia Tang, and Langton Mukwereza. 2016. "Science, technology, and the politics of knowledge: The case of China's agricultural technology demonstration centers in Africa." *World Development* 81: 82–91.

5 The Green Revolution and absent women

Introduction

The Green Revolution as we have seen in the preceding chapters has been a story of technology, scale and government. What has been left out is the diversity of farmers who are involved in agriculture and in particular the role of the women farmers who have always had a central role whether in growing or managing particular crops or in running the farm when the men are away, which is increasingly due to off farm work nearby or longer-term migration to bigger centres or even internationally. This chapter argues that ignoring women as farmers in the provision of support services is a missed opportunity for increasing agricultural production and ensuring food security more generally.

Background

The focus of the various iterations of the Green Revolution has been on the production unit; this invariably has been the peasant family farm, assumed to be a homogeneous unit based on a stereotype of the man as the provider and the woman as the carer, with children being cared for by the woman. Agricultural innovation over the past 50 or more years has generally targeted the male farmer and ignored the role of women in the agricultural production processes (Berger, DeLancey, and Mellencamp 1984; Negin et al. 2009; Theriault, Smale, and Haider 2017). This observation is not new. As mentioned earlier, Loveridge (2017) reports that Prime Minister Nehru of India in 1950

> noted that while he had observed many men engaged in trades, he had seen few women in training. He warned the crowd that if 'the other half does not join hands' in work, the nation could not hope to pull itself up from poverty. Nehru's remark foreshadowed a

recurrent criticism of Indian community development programmes that experts would struggle to address.

(p.61)

Indian agriculture, like agriculture everywhere, still has some way to go to recognise women as farmers (Sachs 1996/2018). Ester Boserup's (1970) seminal work on women's role in agriculture was based on her own observations in India and Africa in the mid-1960s, and how women's role although central to production was continually overlooked by researchers and policy makers. Boserup's work led to the Percy Amendment of the US Congress, an early piece of far reaching legislation that mandated the inclusion of women in all development activities, particularly in rural development, the main focus of foreign aid programs at the time. The Percy Amendment was also based to some extent on similar legislation from Sweden in the 1960s (DAC Secretariat 1975; Snyder 1995; World Bank 1994). It also led to the establishment of the Women in Development office in USAID to oversee the women in development approach across the agency. The problem was that this oversight did not happen in any meaningful way, and the office struggled to have its voice heard for at least the next 20 years, with strong institutional resistance being the norm (Kilby 2015; Miazad 2002). Recent DAC reports suggest that USAID still has some way to go in terms of the proportion of its aid budget being spent on projects that focus on gender and women's equality issues, either as a primary or secondary objective, being a low 21 per cent (DAC 2017).

A systematic male bias in agricultural research and in associated programs has the effect of actively disadvantaging women. The reason is that this bias is based on the implicit, if not explicit, assumption of male headship of households and men being the principle cultivator on the family farm (Pinstrup-Andersen and Hazell 1985; Theriault, Smale, and Haider 2017; Tinker 1976; Wu and Ye 2016). The household system is more complex with women, and children for that matter, playing a key role in food production, so that 'gender takes meaning in the context of age, caste, and livelihoods' (Fisher and Carr 2015, p.83). For aid agencies, policy makers, and agricultural researchers 'it is easy simply to assume that shifts in labour time allocation are possible for both men and women and that they have negligible costs for all concerned' (Blackden and Morris-Hughes 1993, p.4). This is despite evidence to the contrary, which finds this assumption to be problematic where women are decision makers and more commonly where there are various forms of joint decision making. The place of the woman in the community and the attributes women have also plays a role:

> Participation for women is further influenced by age, educational level, time, status, and previous membership in organizations, access to assets and resources, organizations' rules of entry, sociocultural norms and enabling environment.
>
> (Ochago 2017, p.3)

It is not total labour availability that is the issue, but the availability of women's labour given the various important roles they play in many of the agricultural processes (Hird-Younger and Simpson 2013; Pinstrup-Andersen and Hazell 1985; Theriault, Smale, and Haider 2017). Common stereotypes and perceived cultural norms are also a constraint. In the case of cultivation there is often a perceived clear division of labour, with men being responsible for ploughing, while women do other agricultural work such as weeding, threshing etc. (De Schutter and Vanloqueren 2011; Nyantakyi-Frimpong and Bezner Kerr 2015; Pinstrup-Andersen and Hazell 1985). The stereotype of men as being universally responsible for ploughing, if it was ever valid, is being challenged, for example by the Green Army program in Kerala, India. In this case, due to the shortage of men from labour migration and the collapse of rice production in the state, the state government has supported the Green Army program whereby women are trained in using farm machinery and contract themselves out as teams of share-croppers to cultivate the otherwise neglected land of men who have migrated (Alex 2013).

Women as cultivators or farm managers is more widespread than what is the expected norm or what is reported in surveys and the like. The percentage of women as farm managers or principle cultivators in their own right is estimated to be between 30 and 40 per cent of the farming cohort, quite a large group to ignore or overlook (Berger, DeLancey, and Mellencamp 1984; Frank 1999). Projects have failed because men were the targets for the introduction of new varieties of crops when in fact it was the women who were responsible for seed selection, and the men did not necessarily share the information from extension sessions (Berger, DeLancey, and Mellencamp 1984, p.41). In China, where women are becoming increasingly the principle cultivator, existing labour exchange practices are breaking down as men cultivators will not exchange their labour with women for the heavier work due to prevailing cultural norms of men not being seen as working for or with women (Wu and Ye 2016). The policy of focusing state research on larger (male led) farms seems to ignore the untapped resource of women's capabilities and existing roles (China State Council 2017). There is also an assumption that women can get jobs as labourers on these larger units and somehow this is 'empowering'. In fact they are often more vulnerable and receive significantly lower wages, such as is the case of commercial agriculture in Rwanda, or little social recognition beyond low wages, such as in cardamom plantations in Nepal (Bigler et al. 2017; Sony et al. 2016). While the increased work opportunities are real, these are invariably more casual and a poor substitute to addressing the issue of access to services and support to women as farmers.

Women and men of the family often operate as separate economic units in some places, and even lend to each other. While the following observation may be from 40 years ago, it is probably as true today as it was then: 'In parts of West Africa, husbands and wives lend each other cash for economic ventures, at interest rates, as one observer remarked, "only slightly less usurious than those of moneylenders"' (Staudt 1979, p.6). The household is a site of 'cooperative conflict' or bargaining, whereby a whole range of decisions and related activities are negotiated, and trade-offs agreed to (Manfre et al. 2013, p.8). If the man receives a benefit, say from government or a donor, that the woman does not get, then the whole exchange relationship is put at risk, and the woman is worse off (Peña, Webb, and Haddad 1996; Manfre et al. 2013).

Likewise, notions of land ownership and how it functions are more complex:

> property right arrangements may be less determined by notions of 'optimum size' of landholding than by relations among groups of producers and by farmers seeking to adjust land to other resource endowments.
>
> (Conway and Barbier 1988, p.667)

While Conway and Barber do not specifically refer to women, it is clear that the relationship between the *de jure* land owner and the principle cultivator, who may be a woman, is more complex (Berger, DeLancey, and Mellencamp 1984; Fonchingong 1999; Pattnaik et al. 2017). While the principle cultivator is the person requiring access to credit for capital as well as extension support, a poor regulatory system that favours the (usually male) land owner, even if a woman is the principle cultivator or farm manager, effectively discriminates against women.

When women have been left out of the changes brought about by the Green Revolution, both in design and implementation, they are actively disadvantaged and made worse off if their own separate economic roles and needs are not recognised and addressed. Programs can 'wreak havoc on the long-standing balance of labour and return within households if resources and opportunities are channelled to men alone' (Staudt 1979, p.6). This situation becomes more acute in places where there are high levels of male migration such as China and elsewhere, where women-headed households and the structural disadvantages they face are not addressed (Flatø, Muttarak, and Pelser 2017; Wu and Ye 2016).

Women's productivity is not static, nor is it unaffected by their lack of access to new technology and male preferences in official programs, to the point that this neglect of women leads to a decline in overall productivity. Again, this is not new so, again from 40 years ago: 'Women's declining productivity was attributed in part to long-standing male preference in program implementation' (Ashby 1981; Staudt 1979, p.10). It is the long-standing norms concerning gender roles that have been incorporated into aid programs that have led to this decline in productivity. Margaret Mead in the 1970s notes:

> The Euro-American tendency to attribute the concern with agricultural production (with food before it leaves the harvest field) to men and to attribute the concern with food after it leaves the harvest field to women led to the dual assumption that scientific agriculture was a male field and scientific food knowledge (food preservation, nutrition, child rearing, and home management) was a female field. This seemed to be a step in the right direction when it first was developed in the United States and initially was spread around the world through technical agricultural schools for males and home economics schools for females. Actually, its effects have been disastrous.
>
> (Mead 1976, p.10)

Even the concept of women headed households can be problematic if based on crude measurements and the simplistic binary that there are women headed households, or there are not. There are issues of identity: using headship as a gender marker can exclude women who are taking major decisions in a nominally male headed household (Ashby 1981; Theriault, Smale, and Haider 2017; Twyman, Muriel, and García 2015). In these cases, both men and women in surveys will report the man as the head, as social norms require, but it is the woman that is the *de facto* head and farm manager. Migration has become a major driver in this process: in the last 50 years there has been the rapid growth in migration both internally and internationally, so that farm production is now supplemented by off-farm income, in order for the family farm to survive. This is not only due to industrialisation but also the de-agrarianisation of rural areas, often due to neo-liberal structural adjustment policies (SAPs), that were popular in the 1980s and 1990, with elements of them still remaining. These policies continue to result in a reduction in investment in extensions services, particularly those targeted at poor, marginal, and women farmers.

One example from the 1980s is an agricultural intensification program in Ghana, part of which was the provision of agricultural inputs to small-scale farmers using introduced hybrid varieties of maize and rice, which showed strong yield increases. However, a SAP from the World Bank and IMF a few years later deregulated input prices, some of which rose 40 per cent, as well as removing guaranteed minimum farm gate prices for the maize and rice. The effect was to collapse the market thus making the HYVs uncompetitive; the program closed shortly afterwards (Nyantakyi-Frimpong and Bezner Kerr 2015). The issue in this case was not only the SAP but the increased labour requirements that the HYV program brought on local farmers, particularly for women. The program was 'insensitive to the resource needs and risk capacity of small farmers' (p.18), and ignored 'an agricultural approach that values farmer knowledge, considers ecological context, and is sensitive to social inequalities, including class and gender-based access to resources' (p.31). A better designed HYV program that looked at the broader context, with supportive policies and a clearer understanding of the labour constraints that women in particular have, may have been more successful.

These migration effects and the associated gender issues discussed above have implications for how Green Revolution technologies are adopted by and adapted to local communities and contexts. In China, agricultural innovation has faltered due to the high level of men's migration to industrial centres (Peng, Tang, and Zou 2009), with little support to the women left behind through extension and other services (China State Council 2017). The policy implication may be to focus on the left-behind women led families, and how to make their lives easier while increasing family farm output. In 1981, Ashby noted that migration itself is highly selective in terms of the attributes of the men who migrate, with 'those men remaining in the agricultural sector being of low levels of ability, skills and education, so that resources of skill and knowledge may tend to be concentrated among women farmers' (Ashby 1981, p.152). Even when there are men remaining on farms they may not be the principle cultivator or farm manager: it may be the woman of the household. This raises the question as to why the idea of working with the principle cultivator or farm manager on a family farm is not more widely accepted. Green Revolution technologies and the associated support systems have generally ignored the migration effect on family farms and assume a constant ready labour supply throughout the year, that the main source of labour is men, and that women are underemployed (Pinstrup-Andersen and Hazell 1985). These assumptions are readily contested in most contexts and are generally not valid.

Women and patriarchy

This perception of women's limited role in the household agricultural system has its origins in a set of patriarchal values that sees women being positioned outside the production system, which is the men's domain, and the woman's domain is domestic (Dewan 2016; Mead 1976; Mead 2013; Sobha 2007). Even in post or 'reformed' Communist societies these patriarchal values have returned (if they were ever truly absent) and are reflected in policy and practice changes to reflect this dyad (Johnson et al. 2016). In both China and Russia dominant patriarchal values are being made more explicit in government policy and practice with women being actively encouraged into domestic roles and to be seen as a reserve labour force to be tapped only when there are surplus jobs available (Holmgren 2013; Ji et al. 2017; Kizenko 2013). Similar attitudes and knowledge hierarchies can be found in peasant and indigenous systems, which of course have their own patriarchal structures, but often treat agriculture in a different way, where women have explicit and very active roles (Eddens 2017; Schneider 2015; Wu and Ye 2016).

The idea of the production system being a distinct element of the household system is quite arbitrary (Blackden and Morris-Hughes 1993). Caroline Moser's categorisation of the 'triple role of women' – productive, reproductive, and community roles (Moser 1989), was important to make a point, but it seems to have taken on a life of its own, and often in a fairly uncritical way (Levy, Taher, and Vouhé 2000). Of course, these three roles apply to men as much as to women, the question is the proportion of time men and women spend on each role. The other issue with such a categorisation is that these roles do not live in distinct boundaries, and categorising of where certain activities lie can be easily contested. To imply the notion that household activities are not productive can be demeaning and only adds to the idea of a gender hierarchy (Chant and Sweetman 2012; Eerdewijk and Davids 2014; Levy, Taher, and Vouhé 2000; Rao 2012,). For example, what about the women who cook meals for the family and then wrap the left-over rice in banana leaves to take to the market:

these types of debates can easily become silly (Dewan 2016; Joshi 2015; Ochago 2017). As Ashby (1981, p.150) notes: such categorisations 'are generalizations abstracted from a complex reality'.

Arguments about who is responsible for food production lead to another arbitrary gendered distinction between food production and family nutrition, and the associated stereotype that women 'do nutrition' and men 'do production' (Negin et al. 2009). There is no doubt that in some places nutrition has declined while food availability has increased as a result of the Green Revolution and how it can favour the production of certain crops has been noted above (Negin et al. 2009; Safilios-Rothschild 1981; Vercillo et al. 2015). The issue is how the choice in the food being grown can provide for a balanced diet (Negin et al. 2009; Orr et al. 2016). Some Green Revolution advocates have recognised the issue of nutrition and have adjusted their 'branding' accordingly. For example, the 'New Alliance for Food Security and Nutrition' seems to do very little about either food security or nutrition, but rather its focus is on introducing agribusiness and large-scale agriculture to the point that small-scale agriculture and marginal (often women) farmers are sidelined. This is effectively putting old wine in new bottles:

> strategies have become narrowed: actions that tackle food availability through increased production and not structural issues of access, ...[and are] built on the assumption that problems of hunger and malnutrition in SSA [Sub-Saharan Africa] can be tackled by more production.
> (Vercillo et al. 2015, p. 6)

Historically, there has been a hazy boundary between what was called 'home economics' up until the 1970s, which focused on women, as distinct from 'agricultural extension', which focused on men. Even in the 1980s, extension targeting women was still mainly about home economics (Berger, DeLancey, and Mellencamp 1984; Manfre et al. 2013). A 1956 report on home economics noted that women were engaged in productive activities by marketing home processed food and small-scale livestock production (Markwell 1956). Even when the definition of the home was extended to include the nearby fields, women's work was still seen as secondary, even if it was identical to the work carried out by men (Rao 2012).

Women's role in the Green Revolution

Increasing the role of women farmers in Green Revolution processes and activities is complex. This is due to local contextual issues, and the role that embedded patriarchal structures play. Ways to work within these structures or, ideally, challenge them need to be found so that men are accepting of change, so as to avoid a backlash. Ochago (2017) has identified several challenges in changing the prevailing patriarchal norms:

i   Access to group processes which are part of agriculture extension but have rules of entry that limit the access of women. This may be, for example, that these demonstration sessions are limited to the 'land owner'.
ii Age can be an issue where younger women producers are excluded as a result of social norms and peer pressures,
iii An implicit requirement is that participants have reasonable literacy levels to be able to access written materials.
iv Meetings are often scheduled at times that suit men and the trainers, rather than women and the 'spare' time they have available.
v   Women's overall access to public spheres such as meeting places, to public services such as extension services and markets; and even to certain technologies, which are seen to be in the male domain are restricted.
vi Women's voices at public meetings and process can be silent, as a result of the patriarchal processes at play.

Overcoming these challenges requires more than a passive 'including women' statement, but an active affirmative program that is directed at women farmers and farm managers.

Gender and agricultural research in the 2000s

Many, if not most, research and development gender strategies revolve around the phrase to 'include women'. In practice this has been shown not to work and is often token, with the most common gender statement being 'fifty per cent of participants will be women'. Even achieving this is often limited by the prevailing social context and the local patriarchal norms (Dewan 2016). I argue that the first step to include women in agricultural research is for these research programs to move away from thinking about 'including women', to a more proactive set of policies and practice. However, setting resources aside to work with program and activities specifically targeting women has the danger of patriarchal push back. The farmer led research from the 1980s and 1990s is a model that might work (Berger, DeLancey, and Mellencamp 1984; Thorburn 2014). Farmer led research trials across Indonesia as part of the Integrated Pest Management (IPM) program of the 1990s provided options for farmers to test and choose what worked best for them. The next step is to include both men and women farmers (together or separately), and allow them to trial different options for cropping and land use that suits their time availability and other preferences, and enable them to adopt or adapt those technologies that best suit their needs. This approach is by necessity time consuming and expensive but it can be very effective as the IPM and farmer led research experiences have shown (Berger, DeLancey, and Mellencamp 1984; Thorburn 2014).

Such approaches would overcome some of the limitations of the current activities and have a clearer inclusion of gender issues, beyond the use of the phrase 'including women'. The key to such an approach is providing options as, not only is the context of regions within countries different, but the context of each household is also different. By bringing research options to farmers, women and men farmers have the choice to adapt to and improve an existing system. This may include growing fodder crops or a second crop that has lower labour demands. This raises the question of targeting extension to women:

> As men increasingly turn to wage labour and migration, women must devote more of their valuable time to productive tasks, in addition to fulfilling their household responsibilities. Women farmers, therefore, may find little time to participate in agricultural meetings, farm demonstrations, and farmer training courses. They may also be reluctant to undertake new farming practices and new crops that demand more of women's labour.
> (Berger, DeLancey, and Mellencamp 1984, p.45)

As noted above, from the community development processes in India in the 1950s, to the Integrated Pest Management systems in Indonesia in the 1990s, extension was critical to success, but in neither case was it targeted to women. Ashby back in 1981 recommended several steps to target women that still have relevance. Extension services for women still focus on their domestic care roles and nutrition but ignore the key role that women undertake in agricultural production.

i   Extension services should be common to men and women but be packaged in a way that recognises the range of activities that make up production, including nutrition and health, which is relevant to

both men and women. This goes beyond having women farmers included in extension work.
ii Some extension services should be specifically aimed at women to cover the issues women have as farmers.
iii A cadre of women extension officers could be put in place to support extension work for women, given that many cultures/societies do not support men working directly with women (Buehren et al. 2017; Manfre et al. 2013; Twyman, Muriel, and García 2015).
iv Training should be tailored to times that suit women and these may be different times to those that suit men.
v Existing women's groups can be a resource and entry point for extensions services. Creating new groups can be counterproductive and be an additional burden on women's time, while existing groups such as religious or savings groups can be tapped to provide extension messages (Manfre et al. 2013; Ochago 2017).
vi Informal networking among women may be stronger than among men, and so can also be tapped into; for example, the conversations that happen at the well. Ashby argues that through this 'gossip' 'the informal power women have in controlling information and opinion is effectively greater than that of men' (Ashby 1981, p.187).

These steps suggest a different approach to extension than is currently in place based on the landowner and therefore men on larger farms.

The fundamental shift that is occurring in agricultural production in several regions has accelerated since the late 1970s when Ashby undertook her work. The rapid growth in men's migration and the consequent increase in left behind women being the primary cultivators calls for new approaches. In Ghana a 'women extension volunteer (WEV) model, a peer-to-peer extension approach, uses community-based female volunteers to increase agricultural information dissemination in rural northern Ghana' (Hird-Younger and Simpson 2013, p.1). The volunteers were able to greatly increase the reach of the official extension officers. While the training in this case was quite basic, it had the advantage that it could be regularly updated with new information. The WEVs each dealt with groups of farmers so there was as much sharing of information among the farmers as from the extension officers. The other advantage is that both the extension volunteer and the woman farmer were both empowered through this process. In India, for example, the Self-Employed Women's Organisation (SEWA) undertakes rural campaigns through group training and farmer field schools, all aimed at its women members. This includes marketing whereby the organisation links groups with 'seed companies, research institutes and marketing organizations... and helped groups to reach the required standards for a nationally recognized quality mark (the AGMARK) for their packaged products such as cumin' (Gale and Freccero 2013, p.7).

The problem is that these examples are invariably the exception rather than the rule, and bureaucratic inertia and patriarchal values inhibit their widespread adoption. While these examples are a good basis on which to build approaches to research, and the evidence is there that they work, more structural approaches are required within research institutions themselves to enable change to occur and a more mainstream recognition of women as farmers.

Conclusion

This chapter has posed the question as to why women farmers are absent in both agricultural research and policy. The answer seems to be that they are not seen as playing a central role and that the household is the focus under a male head, a view that is more often than not incorrect. Another reason is that any new technology will trickle down to women farmers if they need it, and so it is thought that there is no need for a focus on women farmers. As this chapter has discussed, the farming household is complex and women and men have different management roles and often grow different crops, and as a result have different needs in terms of agricultural technology and support. More recently the family farm often has the woman as the farm's manager due to the man being away working elsewhere. Ignoring the role of women as farmers in their own right is a lost opportunity for research and policy practice, and while this chapter has identified a number of steps that can be taken in reaching women farmers, there is still some way to go before these are adopted across the board.

References

Alex, Jiju P. 2013. "Powering the women in agriculture: Lessons on women led farm mechanisation in South India." *The Journal of Agricultural Education and Extension* 19(5): 487–503.

Ashby, Jacqueline. 1981. "New models for agricultural research and extension: the need to integrate women." In *Invisible Farmers: Women and the Crisis in Agriculture*, edited by Barbara C. Lewis, 144–195. Washington: Agency for International Development, Office of Women in Development.

Berger, Marguerite, Virginia DeLancey, and Amy Mellencamp. 1984. *Bridging the Gender Gap in Agricultural Extension*. Washington, DC: International Center for Research on Women

Bigler, C., Michèle Amacker, Chantal Ingabire, and Eliud Birachi (2017). "Rwanda's gendered agricultural transformation: A mixed-method study on the rural labour market, wage gap and care penalty." *Women's Studies International Forum* 64: 17–27.

Blackden, C. Mark, and Elizabeth Morris-Hughes. 1993. "Paradigm Postponed: Gender and Economic Adjustment in Sub-Saharan Africa." In *World Bank Group Archives*, edited by World Bank. Washington: World Bank. Technical department Africa region. Human resources and poverty division Technical Note 13.

Boserup, E. 1970. *Woman's Role in Economic Development*. London: Allen and Unwin.

Buehren, Niklas, Markus P. Goldstein, Ezequiel Molina, and Julia Vaillant. 2017. "The impact of strengthening agricultural extension services: Evidence from Ethiopia." Policy Research working paper; no. WPS 8169; Impact Evaluation series. Washington, DC: World Bank Group.

Chant, Sylvia, and Caroline Sweetman. 2012. "Fixing women or fixing the world? 'Smart economics', efficiency approaches, and gender equality in development." *Gender & Development* 20(3): 517–529.

China State Council. 2017. *China's Annual Agricultural Policy Goals*. No. 1 Document of the CCCPC and the State Council (unofficial translation). Edited by Global Agriculture Information Network. Beijing: USDA Foreign Agriculture Service.

Conway, Gordon R., and Edward B. Barbier. 1988. "After the green revolution: Sustainable and equitable agriculture for development." *Futures* 20(6): 651–670.

DAC. 2017. *Aid in Support of Gender Equality and Women's Empowerment: Statistics based on DAC Members' reporting on the Gender Equality Policy Marker, 2014–2015*. Paris: OECD.

DAC Secretariat. 1975. "The integration of women in development: Paper from the Swedish delegation for the DAC meetings October 10, 1975, October 1," in *OECD Archives* (DAC F33230 DAC correspondents group on women and development). Paris: OECD.

De Schutter, Olivier, and Gaëtan Vanloqueren. 2011. "The new green revolution: How twenty-first-century science can feed the world." *Solutions* 2(4): 33–44.

Dewan, Ritu. 2016. "Contextualising and visibilising gender and work in Rural India: Economic contribution of women in agriculture." *Indian Journal of Agricultural Economics* 71(1): 49.

Eddens, Aaron. 2017. "White science and indigenous maize: The racial logics of the Green Revolution." *The Journal of Peasant Studies*: 1–20.

Eerdewijk, Anouka, and Tine Davids. 2014. "Escaping the mythical beast: Gender mainstreaming reconceptualised." *Journal of International Development* 26(3): 303–316.

Fisher, Monica, and Edward R. Carr. 2015. "The influence of gendered roles and responsibilities on the adoption of technologies that mitigate drought risk: The case of drought-tolerant maize seed in eastern Uganda." *Global Environmental Change* 35: 82–92.

Flatø, Martin, Raya Muttarak, and André Pelser. 2017. "Women, weather, and woes: The triangular dynamics of female-headed households, economic vulnerability, and climate variability in South Africa." *World Development* 90: 41–62.

Fonchingong, Charles. 1999. "Structural adjustment, women, and agriculture in Cameroon." *Gender & Development* 7(3): 73–79.

Frank, Emily. 1999. "Gender, agricultural development and food security in Amhara, Ethiopia: The contested identity of women farmers in Ethiopia." In *USAID/Ethiopia Project 663–0510*. Addis Abeba: USAID.

Gale, Chris Kathleen Collett, and Piera Freccero. 2013. "Delivering Extension Services through Effective and Inclusive Women's Groups: The Case of SEWA in India." In *MEAS Case Study # 5*. Washington DC: USAID Feed The Future.

Hird-Younger, Miriam, and B. Simpson. 2013. "Women extension volunteers: An extension approach for female farmers." In *MEAS Case Study No. 2*. Washington DC: USAID Feed The Future.

Holmgren, Beth. 2013. "Toward an understanding of gendered agency in contemporary Russia." *Signs: Journal of Women in Culture and Society* 38(3): 535–542.

Ji, Yingchun, Xiaogang Wu, Shengwei Sun, and Guangye He. 2017. "Unequal Care, unequal work: Toward a more comprehensive understanding of gender inequality in post-reform urban China." *Sex Roles* 77(11–12): 765–778.

Johnson, Nancy L., Chiara Kovarik, Ruth Meinzen-Dick, Jemimah Njuki, and Agnes Quisumbing. 2016. "Gender, assets, and agricultural development: Lessons from eight projects." *World Development* 83: 295–311.

Joshi, Deepa. 2015. "Gender change in the globalization of agriculture?" *Peace Review* 27(2): 165–174.

Kilby, Patrick. 2015. *NGOs and Political Change. A History of the Australian Council for International Development*. Canberra: ANU Press.

Kizenko, Nadieszda. 2013. "Feminized patriarchy? Orthodoxy and gender in post-Soviet Russia." *Signs: Journal of Women in Culture and Society* 38(3): 595–621.

Levy, Caren, Nadia Taher, and Claudy Vouhé. 2000. "Addressing men and masculinities in GAD1." *IDS Bulletin* 31(2): 86–96.

Loveridge, Jack. 2017. "Between hunger and growth: Pursuing rural development in Partition's aftermath, 1947–1957." *Contemporary South Asia* 25(1): 56–69.

Manfre, Cristina, Deborah Rubin, Andrea Allen, Gale Summerfield, Kathleen Colverson, and Mercy Akeredolu. 2013. "Reducing the gender gap in agricultural extension and advisory services." In *MEAS Discussion Paper Series on Good Practices and Best Fit Approaches in Extension and Advisory Service Provision MEAS: Discussion Paper 2*. Washington, DC: MEAS.

Markwell, Rachel. 1956. "Agricultural Home Economics Extension Work. Greece: February 15, 1950 – June 1956". Athens: US Mission to Greece.

Mead, Margaret. 1976. "A comment on the role of women in agriculture." In *Women and World Development*, edited by Michele Bo Bramsen, Irene Tinker, and Mayra Buvinid, 9–11. New York: Praeger.

Miazad, Ossai. 2002. "The Global Action and Investments for Success for Women and Girls (GAINS) Act." *Human Rights Brief* 9(3): 37.

Moser, Caroline. 1989. "Gender planning in the Third World: Meeting practical and strategic gender needs." *World Development* 17(11): 1799–1825.

Negin, Joel, Roseline Remans, Susan Karuti, and Jessica C. Fanzo. 2009. "Integrating a broader notion of food security and gender empowerment into the African Green Revolution." *Food Security* 1(3): 351–360.

Nyantakyi-Frimpong, Hanson, and Rachel Bezner Kerr. 2015. "A political ecology of high-input agriculture in northern Ghana." *African Geographical Review* 34(1): 13–35.

Ochago, Robert. 2017. "Barriers to women's participation in coffee pest management learning groups in Mt Elgon Region, Uganda." *Cogent Food & Agriculture* 3(1): 1358338.

Orr, Alastair, Sabine Homann Kee-Tui, Takujii Tsusaka, Harry Msere, Thabani Dube, and Trinity Senda. 2016. "Are there 'women's crops'? A new tool for gender and agriculture." *Development in Practice* 26(8): 984–997.

Patel, Raj. 2013. "The long green revolution." *The Journal of Peasant Studies* 40(1): 1–63.

Pattnaik, Itishree, Kuntala Lahiri-Dutt, Stewart Lockie, and Bill Pritchard. 2017. "The feminization of agriculture or the feminization of agrarian distress? Tracking the trajectory of women in agriculture in India." *Journal of the Asia Pacific Economy*: 1–18.

Peña, Christine, Patrick Webb, and Lawrence Haddad. 1996. *Women's Economic Advancement through Agricultural Change: A Review of Donor Experience*. Washington, DC: International Food Policy Research Institute, February.

Peng, Shaobing, Qiyuan Tang, and Yingbin Zou. 2009. "Current status and challenges of rice production in China." *Plant Production Science* 12(1): 3–8.

Pinstrup-Andersen, Per, and Peter B.R. Hazell. 1985. "The impact of the Green Revolution and prospects for the future." *Food Reviews International* 1(1): 1–25.

Rao, Nitya. 2012. "Male 'providers' and female 'housewives': A gendered co-performance in rural north India." *Development and Change* 43(5): 1025–1048.

Sachs, Carolyn E. 1996/2018 *Gendered Fields: Rural Women, Agriculture, and Environment*. Abingdon: Routledge.

Safilios-Rothschild, Constantina. 1981. "The role of women in modernizing agricultural systems." Paper presented at the Population Association of America Annual Meeting March 26–28, Washington DC.

Schneider, Mindi. 2015. "What, then, is a Chinese peasant? Nongmin discourses and agroindustrialization in contemporary China." *Agriculture and Human Values* 32(2): 331–346.

Snyder, M. 1995. "The politics of women and development." In *Women, Politics and the United Nations*, edited by A. Winslow. Westport Connecticut: Greenwood Press.

Sobha, I. 2007. "Green revolution: Impact on gender." *Journal of Human Ecology* 22(2): 107–113.

Sony, K.C., Bishnu Raj Upreti, Bashu Prasad Subedi, 2016. "'We know the taste of sugar because of cardamom production': Links among commercial cardamom farming, women's involvement in production and the feminization of poverty." *Journal of International Women's Studies* 18(1): 181–207.

Staudt, Kathleen A. 1979. "Tracing sex differentiation in donor agricultural programs." Paper prepared for the American Political Science Assoc. Annual Meeting, Aug. 31 to Sept 3, Washington, DC.

Theriault, Veronique, Melinda Smale, and Hamza Haider. 2017. "How does gender affect sustainable intensification of cereal production in the West African Sahel? Evidence from Burkina Faso." *World Development* 92: 177–191.

Thorburn, Craig. 2014. "Empire Strikes Back: The making and unmaking of Indonesia's national integrated pest management program." *Agroecology and Sustainable Food Systems* 38(1): 3–24.

Tinker, Irene. 1976. "Introduction: The seminar on women in development." In *Women and World Development*, edited by Irene Tinker, Michelle Bo-Bramson and M. Buvinic, 1–8. New York: Praeger.

Twyman, Jennifer, Juliana Muriel, and María Alejandra García. 2015. "Identifying women farmers: Informal gender norms as institutional barriers to recognizing women's contributions to agriculture." *Journal of Gender, Agriculture and Food Security* 1(2): 1–17.

Vercillo, Siera, Vincent Z. Kuuire, Frederick Ato Armah, and Isaac Luginaah. 2015. "Does the New Alliance for Food Security and Nutrition impose biotechnology on smallholder farmers in Africa?" *Global Bioethics* 26(1): 1–13.

World Bank. 1994. "Gender issues in bank lending: An overview." In *Report No. 13246 Operation Evaluation Department*, edited by World Bank. Washington: World Bank Group Archives.

Wu, Huifang, and Jingzhong Ye. 2016. "Hollow lives: Women left behind in rural China." *Journal of agrarian change* 16(1): 50–69.

6 Conclusion

The lessons learned from the Mexico, India, and the China experiences are the need for a more integrated approach in agricultural research, and more importantly to address gender issues and the importance that women farmers play as primary cultivators and farm managers. Plant breeding itself can only lead to increases in productivity if it is complemented by a set of other changes to reflect local context and systems. This variation in the broader policy context explains why the Green Revolution was successful in some countries and not others. National support systems of agriculture infrastructure and support services are necessary to encourage technological change. The post-war Green Revolution with its emphasis on plant breeding in specialist centres tended to assume the required supporting infrastructure would be in place to assist the new technology. Success depended on a set of government and donor policies that were often ignorant of the plant breeding advances, and the role complementary policies played. The lesson came from Europe, the US and Australia in the nineteenth century where government support for land reform, agricultural extension, and technological innovation to increase food production were mechanisms critical to these successes.

Likewise, the success story of China and to a lesser extent India for a time, was that systems and policies were put in place to complement the technical changes. These included land reform, investment in irrigation, fertilizer subsidies, and credit and agricultural extension policies. While it was possible for India and China to assert their sovereignty in setting agricultural policy independent of donors, smaller countries were very much at the mercy of the international aid donors more generally, and the World Bank in particular. From the mid-1970s Western donors including the World Bank were using their aid programs to move away from state interventions in agricultural policy to more free market models, even in contexts where there were few market opportunities for private investment and capital. In addition to the production systems, there were broader livelihood systems where migration and off farm income played a critical role in household well-being. These broader livelihood systems have implications for how new technology and plant varieties are taken up, and what changes are needed in local and national institutional frameworks to support their uptake.

The twenty-first-century Green Revolution with its strong focus on Africa is repeating some of the mistakes of the past; that is, a focus on imported technology, particularly HYVs from either China or the West into alien environments, without a full understanding of the local socio-political contexts, or even the differing agro-environmental contexts. If these mistakes from Asia are to be avoided, investment is required in agricultural research at the farming systems level. As I have discussed, this has been attempted sporadically from the late nineteenth century through to the 1980s and 1990s, and can be found in some of the work of the USAID supported Feed the Future Program; the issue, however, is that this has not been sustained due to ideological positions held by the major players. The challenge is the rivalry between China and the West with not only competing infrastructure development, such as the Belt Road Initiative from China, but a more commercial based agricultural research program. Here large agro-businesses with their use of high yielding hybrid and genetically modified varieties of crop with large capital inputs are often favoured over the small-holder and women farmer in complex farming systems that are less capital intensive and more risk averse, but with huge opportunities for increased yields with smaller scale research input.

A closely related issue is something that has applied through the long history of agricultural technological change and the Green Revolution in all its iterations; this is the lack of appreciation of the role of gender relations and the role women play in agricultural production systems. This lack of appreciation has resulted in not only missed opportunities but also a loss of potential productivity as women are increasingly the principle cultivator and farm manager in complex household systems where the men migrate. The solution as I have suggested in this volume is a change in the way agriculture research is conducted to a more systems approach in which it is the farm manager or principal cultivator, rather than the land holder, who should be the focus of attention.

Index

Africa Agricultural Development 46
African Green Revolution (AGRA) 47–50
agricultural research 42, 57, 72, 73; agricultural demonstration farms/centres 44; and gender 44, 57, 64, 67; foreign aid funded 57; foundations led 3; state led 8, 34, 37
agricultural support 16, 42
agricultural: extension 8, 14, 26–29, 33, 42, 58–61, 64–66, 72; productivity 8, 14, 16, 28, 33, 35, 42, 44, 60, 72, 73; support 16, 42; transformation 24
agro-climatic 11, 26, 28, 73
aid (see development and aid)
Annan, Kofi 47
Aswan High Dam (see dams)
authoritarian 14, 17

Belt Road Initiative (see China)
Black, Eugene 16
Borlaug, Norman 1, 4, 12, 13, 17, 25, 31
Boserup, Ester 57
Bräutigam, Deborah 15, 43–46

capital (availability) 7, 12, 19, 26, 28, 35, 47, 59, 73
capitalism (system) 26, 33, 50
China: and Africa 3, 14, 15, 44, 45; Belt Road Initiative 16, 73; Chinese Agricultural Program (Africa) 3, 14–16, 45, 46, 48, 50; Chinese owned farms (in Africa) 15, 45; Cold War 9, 13, 14; eight principles of foreign aid 16; government 16, 36, 48; Great Leap Forward 34, 35, 37; Green Revolution 1, 3, 6, 8, 9, 14, 16, 24, 32–36, 42, 43, 46, 51, 72; high yielding varieties 3, 12, 35, 50, 73; irrigation 3, 17, 35; 'Red China' 9, 13, 14; State Council 36, 58, 61; women farmers 36, 58, 60–62, 72
Cold War 6, 9,12–14, 16–18, 43
Communes 34
Communism 3, 4, 9, 11, 12, 13, 30
Communist, anti (see Communism)
community development program (India) 28–30, 57
community roles (see women)
Congress (US) 9, 57
cooperatives (see farming systems)
corporate farming (see farming)
Country Cooperation Frameworks 48
credit (inputs) 7, 26, 32, 33, 51, 59, 72
crops: maize, 1, 4, 8–11, 25–28, 46, 61; millet 10, 43; rice 1, 9, 10, 14, 15, 32, 51, 58, 61, 62; rice (HYV) 4, 9, 35, 43, 46, 51; sorghum 10, 44; wheat 1, 10, 12, 25, 27, 28, 32, 37, 43, 51; wheat (HYV) 4, 8, 9, 24–26, 31, 32, 43; wheat (rust resistant) 8, 25
CRSP (Cooperative Research Support Program) 48
culture: cultural context 16, 30; cultural norms 58
customary title (see indigenous)

CYMMIT (International Maize and Wheat Improvement Center) 9

DAC (Development Assistance Committee) 13, 57
dams 16–18; Aswan High Dam 17; as propaganda 17; large 3, 7, 16, 17
development: assistance (foreign aid) 9, 10, 12, 14–16, 30, 42, 43, 45, 47, 57; agricultural (rural) 1, 3, 9, 15, 16,28, 34, 43–46, 49, 52, 57; grassroots (see community led development); industrial 16, 28, 43, 46, 73; socio-economic 1, 13; sustainable (see sustainable development); technical 46; and women 57, 64
Development Assistance Committee (see DAC)
donors 1, 15, 17–19, 45, 59; policies 15, 42, 43, 44, 72
dwarf varieties (HYV) 25, 26

economic theory: communist 17; neo-liberal 33, 42, 61; New Structural Economics 44
eight principles of foreign aid (see China)
ejidos (see Mexican)
Enlai, Zhou 16
education 16, 18, 53, 49, 58, 61
exchange: labour 58, 59; relationships 44
extension (see agriculture)

family farms (see farming)
farm inputs (see inputs)
farm managers (see women farmers)
farmer demonstration (centres) 14, 25, 31, 44–46, 64, 65
farming: collective farms 6, 34; corporate (interests) 2, 15, 44, 47, 48, 51; family farm 2, 6–8, 35, 36, 44, 51, 52, 56, 57, 60, 61, 67; larger farms 2, 6, 10, 13, 24, 26, 30, 32, 33, 36, 37, 47, 58, 66; peasant 2–4, 6–8, 10, 24–29, 32–34, 36, 42, 47, 50, 52, 56, 62; smallholder (small-scale) 6, 24, 26, 27, 31, 32, 37, 47, 63, 73
farming systems: capital intensive, 73; local 2; research 2, 29, 42, 48, 73
Farrer, William 8
Feed the Future program 48, 50, 73
feudal 28, 29, 33, 34, 37
FOCAC (Forum on China Africa Cooperation) 44, 46, 48
food: aid 12, 30, 31; availability 7, 31, 63; distribution 4, 13, 25, 30, 33, 34, 51; prices 7, 31, 33; production 1, 4, 7, 8, 12, 14, 28, 35, 47, 48, 58, 63, 72; security 11, 31, 44, 45, 47, 50, 56, 63
Ford Foundation 28
foreign aid (see development assistance)
foreign exchange 18, 28, 31
Forum on China Africa Cooperation (see FOCAC)
free market 14, 31, 72

Gandhi, Indira 31, 33
Gates Foundation (Bill and Melinda) 47
Gaud, William 9
gender: equality 44, 57, 61; policy 36, 64, 65; relations 3, 52, 60, 61, 73; roles 60–63
geo-political 10, 29, 30
Germany 8
Global South 10, 12, 13, 16, 30
GMOs (see high yielding varieties)
Great Leap Forward (see China)
Green Revolution: as a metaphor 9; distributional effects 33; Europe (1870–1920) 3, 4, 6, 7, 10, 32, 51, 72; liberal based 33, 34; post-war (1940–1980) 3, 6, 9, 11, 46, 72
Group of Eight (industrialised countries) 47

Harrar, George 4, 25
health 16, 47, 49, 51, 65
high yielding varieties (HYVs) 3, 17, 28, 30, 32, 35, 43, 44, 48, 49, 73; GMO 43, 47, 50, 51; hybrid 8, 11, 34–36, 43, 46, 47, 50, 61, 73; maize 26, 27, 61; rice 32, 43, 46, 61;

wheat 4, 8–10, 12, 24–28, 31, 32, 43
home economics 60, 63
household: head 57, 60, 67; production 35, 36; system 51, 57, 59, 62, 65, 67, 73; women headed 60, 61
hunger 14, 63
hybrid (see HYV)

India 8–14, 17, 19, 24, 28–33, 37, 43, 47, 51, 56–59, 65, 66, 72
indigenous 10, 62; customary title 15; Mexican native Americans 10, 11; Mayan Indians 11
Industrial Revolution 6
inputs 16; farm 7, 10, 14, 16, 29, 32, 49, 61, 73; fertilizer 28; irrigation 33; seeds 34, 51
investment: China 14, 44, 45; farm level 19, 31, 36, 73; infrastructure 3, 16–19, 44; private 15, 16, 73; state 1, 34, 36, 42, 43, 61, 72; US 30, 31
IRRI (International Rice Research Institute) 9
irrigation 1, 3, 6, 14, 16–18, 24, 33–35, 37, 48, 72

Johnson, Lyndon 31, 32

labour (see rural)
Land Army 8
land grant universities 8
land ownership (see land tenure)
Land reform (see land tenure)
land tenure: distribution 37; land grabs 44; reform (land reform) 1, 7, 8, 14, 24, 26, 28, 30, 34, 35, 42, 72
landlords (see feudal)
larger farms (see farms)
local knowledge 11

maize (see farms; crops)
male bias: in agricultural research 57; as household heads 36, 57, 59–62, 67
malnutrition 10, 63
marginalisation: social 2, 44; of women 1, 2

marketing 7, 9, 63, 66
Mead, Margaret 60
Mechanics Institutes 8
Mexican Agricultural Program (MAP) 10, 25
Mexican native Americans (see indigenous)
micronutrients 51
migration 36, 51, 73; and men 36, 56, 58, 60–62, 64, 66
Millennium Partnership for Africa (MPA) 44
Millennium Villages Program (MVP) 3, 48–50
millett (see crops)
model villages (see Millennium Villages Program (MVP))
modernisation 3, 6

New Alliance for Food Security and Nutrition (NAFSN) 47–50
natural resources 29
Nehru, Jawaharlal 28–31, 56
neo-liberal (see economic theory)
New Economic Partnership for African Development (NEPAD) 44, 46
New Structural Economics (see economic theory)
nutrition 4, 10, 11, 47–51, 60, 63, 65

oil crisis 33

Paddock, William and Paul 13
patriarchy 9, 62
peasant (see farming)
Percy Amendment 57
plant breeding 2–4, 10–12, 14, 17, 32, 72
policy: agricultural 2, 8, 9, 13, 14, 26–28, 30, 31, 34–37, 42, 43, 72; extension; foreign 9, 12, 33; gender (see gender); policy-makers 27, 57; research 2, 37, 58, 67
population 7, 13, 28, 30, 32, 35, 50; growth 13, 30, 35
poverty 10, 26, 50, 56; alleviation 10
principle cultivator (see women farmers)
propaganda 17

reforms (economic, political) 31, 33
research, agricultural (see agricultural research)
risk 4, 18, 59; averse 11, 27, 73; farming 2, 27, 32, 48, 49, 51, 61
Rockefeller: brothers 1, 3, 6, 10, 12, 19; Nelson 3, 4, 9, 25; John D. 9, 13; Foundation 3, 10–13, 19, 25, 30–32, 34, 35, 47, 48
rural: infrastructure 16–18, 24, 26, 33, 34, 36, 42, 44, 45, 51, 72; labour 7, 8, 29, 33, 36, 44, 51, 52, 57–62, 65
rust resistant (see crops, wheat)

Sachs, Geoffrey 48
self-sufficiency (food) (see nutrition)
'settler' countries 7
smallholder (see farmers)
social dislocation 2, 26
socio-cultural context (see culture)
sorghum (see crops)
sovereignty 31, 72
Soviet (bloc) 9, 12, 16, 17, 30, 34, 49
state: expenditure 42; intervention 8, 49, 72; investment 28, 34, 51, 58; -owned farms 45; support 6, 14, 24, 33, 36
structural adjustment 42, 43, 61,
Subramaniam, Chidambaram 29, 31
sustainable development 13
Sustainable Development Goals (SDGs) 49

technocratic approach 16, 19, 45
technology (agricultural) 2, 8–10, 12, 14, 19, 28, 30, 33, 45–48, 51, 56, 60, 67, 72, 73
tenants (serfs) 30
Tennessee Valley Authority (TVA) 16–18, 35
tenure, land (see land tenure)
'transpatriarchies' (see patriarchy)
Truman, Harry 9, 11

USAID 9, 29, 57, 73

viability (economic) 18

Wallace, Henry 3, 25
welfare 16, 36
Wellhausen, Edwin J. 10, 11, 32
West, the 1, 5, 6, 8, 11–14, 16, 17, 19, 42, 43, 45–47, 73
Western: donors 1, 42, 44, 72; technology 10, 12, 19, 47
wheat (imports) 31, 32
women: community roles 62; disadvantage 59, 60; productiv(ity)/reproductive 36, 60, 62, 63, 65, 72, 73; support services for 56, 72
women farmers: extension to 58, 59, 61, 63–66; as farm managers 37, 58, 64, 72; as principle cultivator 2, 3, 37, 48, 51, 57–59, 61, 73
women headed households (see households)
Women in Development Office (see USAID)
World Bank 1, 3, 15–19, 42, 43, 61, 72